Flammability Handbook for Plastics

Third Edition

CARLOS J. HILADO
PRODUCT SAFETY CORPORATION

Flammability Handbook for Plastics

a TECHNOMIC® publication

265 POST ROAD WEST, WESTPORT, CT.
WESTPORT, CONNECTICUT 06880

LCC-74-82519
ISBN 087762-306-6
Printed in USA

INTRODUCTION

The general class of materials known as plastics has become the most versatile and widely useful class of materials known to man. The range of processing and performance characteristics, and of combinations of these characteristics, made available through judicious design has brought these materials into almost every conceivable application. Many of these applications involve a significant possibility of exposure to fire.

A plastic is defined as a material which contains as an essential ingredient an organic substance of large molecular weight, is sold in its finished state, and at some stage in its manufacture or its processing into finished articles, can be shaped by flow. Because the essential ingredient of a plastic material is an organic substance, combustion can result under sufficiently severe exposure to heat and oxygen. The severity of exposure required to produce combustion, and the results of the combustion process, vary as widely as the materials themselves. Since overall fire hazard is a function,not only of the fire performance of the material, but also of the degree of exposure of the material, and the probability of occurrence and the nature of the fire involved, all of these factors must be considered in determining the degree of inherent fire resistance which the material must possess to be satisfactory for a particular application.

Considerable effort has gone into the study of various aspects of flammability, so that these materials which are proving so useful to man will always be used in ways that will not compromise his safety. The task will be a continuing one, because the family of plastics is growing, and, along with it, its variety of applications; some of these future applications can not even be conceived of at the present time. The needs of man and his society are also changing, and with them the factors that affect his safety, comfort, and convenience.

Because a flammability handbook for plastics must necessarily involve a variety of sciences and technologies spread across the whole spectrum of human knowledge, it is impossible to go into all the subjects in great depth, and any details extracted for attention are brought out because they are believed to be significant to the overall effort to make plastics as useful and as safe as humanly possible.

To Patty

ACKNOWLEDGEMENT

The author is grateful to Mr. William A. Ashe of BASF Wyandotte Corporation, Mr. A. Briber of Underwriters Laboratories, Mr. Gordon H. Damant of the California Bureau of Home Furnishings, Dr. Stanley B. Martin of SRI International, Dr. Gordon L. Nelson of the General Electric Company, Dr. Roanald V. Petrella of the Dow Chemical Company, and Dr. Edwin E. Smith of the Ohio State University, for their assistance in studying this vast field of knowledge.

The author is indebted to Miss Patricia A. Huttlinger, without whose assistance the Product Safety Corporation might never have been established, and without whose encouragement and support this book might never have been written.

FLAMMABILITY HANDBOOK FOR PLASTICS
THIRD EDITION

TABLE OF CONTENTS

CHAPTER 1

Materials for the Plastics Industry

Plastic materials fall into the general class of materials known as synthetic polymers: organic substances of high molecular weight, manmade from repeating units of lesser molecular weight called monomers.

SECTION 1.1. PHYSICAL CLASSIFICATION

Plastics or polymers are available in a variety of physical shapes. The physical form in which a plastic is present has signficant influence on its flammability characteristics. In some cases physical structure is much more important than chemical structure where fire behavior is concerned. For this reason, physical classifications will be discussed first. Six basic physical groupings will be considered: formed, filled, film, foam, fiber, and fines. These groupings are listed in approximate order of increasing susceptibility to flame.

1. **Formed,** the general grouping which includes rods, blocks, castings, extrusions, and moldings, covers structures in which there is a significant thickness and mass of polymer, so that heat transfer to and heat absorption by adjacent material is a function of the characteristics of the polymer itself. Formed plastics generally have the lowest surface to volume ratio and tend to be the least flammable of the physical classifications.

2. **Filled,** the general grouping which includes reinforced plastics, covers structures in which solid materials other than the polymer under

discussion are more or less uniformly distributed throughout the material, to a large enough extent that heat transfer to and heat absorption by adjacent material is a function of both the characteristics of the polymer and the characteristics of the filler or reinforcing material.

3. **Film,** the general grouping which includes films, sheeting, coatings, and adhesives, covers structures in which thickness is very small relative to length and width. Film is strictly defined as sheeting having nominal thicknesses not greater than 0.010 inch, but the term is used loosely here to simplify classification. The small thickness results in relatively little mass per unit area, so that the material is essentially exposed or unexposed surface, and heat transfer to and heat absorption by adjacent material is essentially a function of the characteristics of the adjacent material. In the case of adhesives, solids are the adjacent materials. In the case of coatings, the adjacent materials are a solid and air. In the case of self-supporting sheeting, air is the adjacent material on both sides.

4. **Foam,** the general grouping which includes cellular plastics and honeycomb structures, covers structures in which gaseous materials are more or less uniformly distributed throughout the polymer. These are characterized by a relatively low mass per unit volume, high surface area per unit volume, and low thermal conductivity. Heat transfer to and heat absorption by adjacent material is a function of the characteristics of the polymer, the characteristics of the contained gas, and the manner of distribution of polymer and gas.

5. **Fiber,** the general grouping which includes fabrics and textiles, covers structures in which the polymer is present in the form of essentially continuous cylindrical rods of very small diameter, arranged by weaving or other methods into more or less uniform matrices of varying characteristics. Heat transfer to and heat absorption by adjacent material is a function of the characteristics of the polymer and the characteristics of the matrix, air being the intervening material in both cases.

6. **Fines,** the general grouping which includes powders and dusts, covers materials which are finely divided into small particles and are characterized by high surface area per unit volume or unit mass, and rather complete exposure to the surrounding atmosphere. Fines are the most susceptible of the physical classifications and under certain conditions can produce dust explosions.

SECTION 1.2. CHEMICAL CLASSIFICATION

With the preceding background, a discussion of the various chemical classes of polymers becomes more meaningful from the application viewpoint. Polymers have traditionally been divided into two general groups: thermoplastic and thermosetting. Thermoplastic polymers are capable of being repeatedly softened by increase of temperature and hardened by decrease of temperature, the change upon heating being physical rather than chemical. Thermosetting polymers are capable of being changed into a substantially infusible or insoluble product when cured by application of heat or chemical means.

The major types of thermoplastic materials are olefin (including polyethylene, polypropylene, and fluoroplastics); vinyl (including polyvinyl chloride, polystyrene, acrylonitrile-butadiene-styrene (ABS), and acrylic); cellulosic (including cellulose nitrate, acetate, butyrate, and propionate); and aromatic (including polycarbonate, phenoxy, polysulfone, and polyimide). Other thermoplastic materials are polyamide (nylon) and acetal.

The major types of thermosetting materials are polyurethane; phenolic (including phenol-formaldehyde); amino (melamine-formaldehyde and urea-formaldehyde); epoxy; and polyester.

Sales of plastics in the United States are shown in Table 1.1.

SECTION 1.3. OLEFIN POLYMERS

Polyethylene is produced commercially from ethylene monomer and is divided into two general types: high-pressure or low-density polyethylene, and low-pressure or high-density polyethylene. There is a wide variation feasible in physical properties and processing characteristics, depending on the molecular weight distribution, extent and type of branching, crosslinking, and crystallinity.

The wide variety of applications which can result from this versatility is illustrated by the following list of examples: thin-walled containers by thermoforming, injection molding, or blow molding; pipe; wire coating; paper coatings; floor polishes; textile treatment; films and sheeeting; blister packaging; and toys.

Total polyethylene sales were 5,303,000 metric tons in 1980 (Tables 1.2 and 1.3).

Polypropylene is produced commercially from propylene monomer. It has the lowest density of commercially available thermoplastics, and exhibits high yield strength and rigidity, exceptional flex life, good surface hardness, resistance to most chemicals, and excellent dielectric properties. Applications include interior trim parts for automobiles, bucket seat

Table 1.1. Sales of Plastics in the United States

	Sales, 1000 metric tons	
	1979	1980
Polyethylene, low density	3,552	3,350
Polyvinyl chloride and copolymers	2,790	2,469
Polyethylene, high density	2,224	1,953
Polypropylene and copolymers	1,770	1,647
Polystyrene	1,817	1,608
Polyurethane foams	856	744
Phenolic	788	701
Urea and melamine	700	589
Polyester, thermoplastics (PBT, PET)	409	481
Acrylonitrile-butadiene-styrene (ABS)	563	441
Polyester, thermoset	524	440
Other styrenes	410	349
Acrylic	268	235
Alkyd	225	210
Thermoplastic elastomers	139	154
Other vinyls	163	150
Epoxy	150	144
Nylon	144	125
Polycarbonate	107	99
Modified phenylene oxide	60	53
Cellulosics	65	51
Styrene-acrylonitrile (SAN)	56	48
Polyacetal	48	43
Others	40	33
Total	17,868	16,117

backs, heater ducts, dishwasher racks, ice cube trays, television cabinets and components, wire coating, coaxial cable coating, fibers, pipe, hospital equipment, housewares, luggage, film, shrink packaging, decorative ribbon, brush bristles, and indoor-outdoor carpeting.

Total polypropylene sales were 1,647,000 metric tons in 1980 (Table 1.4).

Polybutylene is produced commercially by polymerization of butylene. It exhibits low cold flow, flexibility, resistance to stress cracking, and chemical stability. Applications include well pipe, water service tubing, collapsible irrigation tubing, and heavy duty packaging.

Polyallomers are a class of olefin block copolymers which exhibit higher

Table 1.2. Uses of Low Density Polyethylene

	Consumption 1000 metric tons	
	1979	**1980**
Extrusion film (12 mil and under)	1,983	1,795
Injection molding	295	236
Extrusion coating	245	236
Extrusion wire and cable	193	161
Rotomolding	34	32
Other extrusion	36	28
Blow molding	27	27
Extrusion pipe and conduit	15	9
Extrusion sheet (over 12 mil)	12	8
Export	408	516
Other (resold, blending, compounding, etc.)	304	302
Total	3,552	3,350
Packaging film	1,168	1,084
Nonpackaging film	877	845
Less film from resold resin	−62	−134
Total extrusion film	1,983	1,795
Paperboard	141	136
Paper	52	50
Foil	27	26
Film	25	24
Total extrusion coating	245	236

crystallinity than earlier copolymers of the olefins. Commercially available polyallomers are of the propylene-ethylene type, and offer high stiffness and high impact, exceptional flow properties, and resistance to flexing fatigue. Applications include heat-sterilizable containers, typewriter cases, integral hinge applications, and cowl panels.

Ethylene-vinyl acetate copolymers are produced by polymerizing ethylene with vinyl acetate, and exhibit excellent low temperature flexibility, high impact strength, low elastic moduli, good resilience, good clarity, and dielectric sealability. Applications include medical tubing, disposable syringes, construction film, shower curtains, cattle ear tags, pool table bumpers, and automobile mudflaps.

Ionomers are ion-linked ethylene-methacrylic acid interpolymers. They are characterized by toughness, flexibility, high transparency, and chemical resistance. Applications include housewares, toys, safety

Table 1.3. Uses of High Density Polyethylene

	Consumption 1000 metric tons	
	1979	**1980**
Blow molding	784	705
Injection molding	488	427
Extrusion	459	395
Rotomolding	26	21
Export	277	232
Other (resold, blending, compounding, etc.)	190	173
Total	2,224	1,953
Household chemicals bottles	290	264
Milk bottles	236	215
Pharmaceuticals/cosmetics bottles	75	66
Other food bottles	68	62
Drums (15 gal. and larger)	30	27
Tight-head pails	27	25
Toys	12	11
Fuel tanks (all types)	14	5
Housewares	5	5
Other blow molding	27	25
Total blow molding	784	705
Industrial pails	94	73
Housewares	68	61
Beverage-bottle bases	29	39
Crates other than dairy, cases, pallets	37	34
Dairy tubs	34	32
Toys	29	26
Caps and closures other than milk bottle	21	20
Ice cream containers	20	19
Dairy crates	18	17
Milk-bottle caps	10	9
Paint cans	7	7
Other food packaging	5	4
Other injection molding	116	86
Total injection molding	488	427
Corrugated pipe	148	127
Wire and cable	57	50
Film for merchandise bags	36	30
Film other than bags and food packaging	37	30

Continued

Table 1.3. Uses of High Density Polyethylene (Continued)

	Consumption 1000 metric tons	
	1979	1980
Pipe other than corrugated, water, and gas	33	29
Water pipe	32	27
Gas pipe	31	26
Film for food packaging	27	25
Sheet (over 12 mil)	30	24
Coating	14	11
Film for deli paper	5	5
Other extrusion	9	11
Total extrusion	459	395

(Concluded)

shields, tool handles, containers, film, sheet, and coatings.

Methylpentene polymer is produced by polymerization of 4-methylpentene-1, and is characterized by extremely low density, high light transmission, high melting point, and excellent electrical properties. Applications include hospital and laboratory ware, encapsulation of electronic relays, piping in milking machines, buttons, and frozen dinner trays.

Chlorinated polyethylene is produced by chemical substitution of chlorine on linear polyethylene. Variations of the molecular weight of the polyethylene feed stock, chlorine content, and chlorine placement produces a wide range of characteristics. Applications include flexible film and sheet, electrical insulation, cable jacketing, and molded goods.

Fluoroplastics are paraffinic hydrocarbon polymers in which the hydrogen atoms have been wholly or partly replaced with fluorine atoms. Chlorine atoms can also be part of the polymer. Polytetrafluoroethylene is a homopolymer of tetrafluoroethylene, while fluorinated ethylene-propylene polymers are considered copolymers of tetrafluoroethylene and hexafluoropropylene. These polymers offer chemical inertness, thermal stability, low friction coefficient, high electrical resistivity, and low dielectric constant. Applications include insulation of aircraft, missile, and spacecraft wiring; insulators for high frequency service; pump linings; packings and gaskets; bearings and seals; piston rings; non-stick coatings for utensils.

Chlorotrifluoroethylene polymer is produced in varying formulations

Table 1.4. Uses of Polypropylene

	Consumption 1000 metric tons	
	1979	**1980**
Extrusion	650	630
Injection molding	554	490
Blow molding	28	26
Export	325	300
Other (resold, blending, etc.)	213	201
Total	1,770	1,647
Fibers and filaments	464	436
Oriented film (up to 10 mil)	96	102
Unoriented film (up to 10 mil)	32	32
Sheet (over 10 mil)	14	17
Straws	14	14
Pipe and conduit	11	11
Coating	5	5
Wire and cable	5	5
Other extrusion	9	8
Total extrusion	650	630
Housewares	77	77
Packaging closures	57	60
Transportation battery cases	64	59
Other transportation products	95	47
Medical	45	45
Packaging containers and lids	39	40
Appliances	45	39
Toys and novelties	32	34
Furniture	19	16
Luggage and cases	5	5
Other injection molding	76	68
Total injection molding	554	490
Consumer packaging	19	17
Medical containers	9	9
Total blow molding	28	26

thermal properties, low moisture absorption and vapor transmission, chemical resistance, and excellent flexibility. Applications include hookup wire insulation, flexible cable jacketing, gaskets, bottle caps for highly corrosive liquids, tubing, and film.

SECTION 1.4. VINYL POLYMERS

Vinyl chloride polymers and copolymers are produced by polymerization of vinyl chloride, with or without co-monomers such as vinyl acetate. Polyvinyl chloride can be prepared by a variety of processes, and can be formulated for numerous applications. These include automobile seat covers, moldings and floor mats, shower curtains, kitchen utensils, baby pants, meat wrap, adhesives, house siding, weatherstripping, flooring, pipe, records, raincoats, toys, paper and textile coatings, gloves, and medical tubing. Vinyl foam is used in seat cushions and marine products such as buoys.

Polyvinyl acetate is prepared by polymerization of vinyl acetate. Polyvinyl alcohol is prepared by hydrolysis of polyvinyl acetate and is used for water-soluble packaging. Polyvinyl butyral is prepared from polyvinyl alcohol and butyraldehyde, and is used in safety glass interlayers and table tops.

Total sales of polyvinyl chloride and its copolymers were 2,469,000 metric tons in 1980 (Table 1.5).

Polystyrene is manufactured by the polymerization of styrene monomer, and exhibits good physical and dielectric properties, chemical resistance, dimensional stability, and versatility in coloring. Polystyrene is modified with rubber to provide impact resistance, and copolymerized with acrylonitrile to improve hardness, rigidity, tensile strength, and chemical resistance. Applications include housewares, toys, closure caps, luggage, appliance housings, automotive parts, lighting diffusers, illuminated interior signs, and, for modified polystyrene, business machine shells, telephones, and packaging. Polystyrene foam applications include building insulation, cushioned shipping containers, and wall panels.

Total sales of polystyrene were 1,608,000 metric tons in 1980 (Table 1.6).

Styrene-butadiene elastomers combine the end-use properties of vulcanized elastomers with the processing advantages of thermoplastics, and are produced from styrene and polybutadiene. They exhibit good resilience, elastic recovery, tensile properties, frictional properties, and chemical resistance. Applications include footwear, toywheels, swim fins, aircraft oxygen masks, stair treads, bumpers, suction cups, and pipe seals.

Acrylonitrile-butadiene-styrene (ABS) polymers are manufactured from the basic building blocks of styrene, acrylonitrile, and polybutadiene. A wide range of performance and processing characteristics can

Table 1.5. Uses of Polyvinyl Chloride

	Consumption 1000 metric tons	
	1979	**1980**
Extrusion	1,619	1,406
Calendering	408	344
Injection Molding	162	132
Dispersion coating	174	128
Dispersion molding	92	78
Compression molding (sound records)	73	52
Blow molding (bottles)	38	45
Solution coating	46	37
Vinyl latexes (adhesives/sealants)	32	27
Export	146	220
Total	2,790	2,469
Water pipe	423	364
Electrical wire and cable	195	177
Conduit	190	164
Sewer/drain pipe	170	132
DWV pipe	138	109
Packaging film	114	103
Irrigation pipe	85	85
Building siding and accessories	36	77
Other pipe and conduit	32	23
Window profiles	24	23
Blood/solution bags	18	18
Garden hose	18	16
Packaging sheet	16	14
Foam moldings	14	14
Weatherstripping	14	14
Medical tubing	12	14
Appliances	14	11
Gas pipe	10	9
Other pressure pipe	10	9
Vehicle floor mats	9	8
Building lighting	8	7
Mobile home skirt	9	5
Transportation bumper strips	5	4
Stationery/novelties	4	4
Gutters/downspouts	1	2
Total extrusion	1,619	1,406

(Continued)

Table 1.5. Uses of Polyvinyl Chloride (Continued)

	Consumption 1000 metric tons	
	1979	**1980**
Building flooring	80	59
Furniture upholstery	55	41
Packaging sheet	32	34
Auto upholstery/trim	41	30
Tablecloths, mats	18	17
Wallcovering	18	16
Pool/pond liners	14	14
Footwear	14	14
Toys	15	13
Building paneling	11	10
Handbags/cases	11	10
Sporting/recreation	10	8
Luggage	9	8
Hospital and healthcare	7	8
Credit cards	7	8
Other transportation upholstery/trim	9	7
Auto tops	9	5
Window shades/blinds/awnings	6	5
Decorative film (adhesive-back)	5	5
Tapes, labels	5	5
Shower curtains	5	4
Waterbed sheet	5	4
Electrical tapes	3	3
Bookbinding	2	2
Stationery/novelties	2	2
Baby pants	2	2
Other building products	2	1
Other calendering	11	9
Total calendering	408	304
Pipe fittings	56	43
Electrical/electronic plugs, connectors	34	27
Footwear	23	18
Electrical appliances, business machines	18	16
Hospital and healthcare	5	7
Transportation bumper parts	7	5
Other building products	3	2
Other injection molding	16	14
Total injection molding	162	132

(Continued)

Table 1.5. Uses of Polyvinyl Chloride (Continued)

	Consumption 1000 metric tons	
	1979	**1980**
Building flooring	84	55
Auto upholstery/trim	14	11
Furniture upholstery	11	9
Carpet backing	9	8
Window shades/blinds/awnings	7	6
Wallcovering	7	6
Apparel/outerwear	7	5
Anticorrosion coatings	5	5
Luggage	5	4
Tablecloths, mats	5	4
Other transportation upholstery/trim	2	2
Hospital/healthcare	2	2
Other dispersion coating	16	11
Total dispersion coating	174	128
Transportation	18	14
Packaging closures	14	13
Sporting/recreation	8	7
Traffic cones	7	7
Footwear	7	6
Handles, grips	7	5
Appliances	7	5
Adhesives	5	5
Sealants	4	4
Toys	3	2
Other dispersion molding	12	10
Total dispersion molding	92	78
Adhesives/coatings	41	33
Packaging cans	5	4
Total solution coating	46	37

(Concluded)

be obtained by varying the composition and the polymerization process. Applications include housings and cases for appliances, refrigerator door and inner liners, luggage, automotive air ducts, housings for lamps and turn signals, pipe and fittings in residential sanitary systems, safety helmets, housings for electronic components, machine gun grips and gun stocks, trays and tote toxes, and boat hulls and decks.

Total sales of ABS were 441,000 metric tons in 1980 (Table 1.7).

Table 1.6. Uses of Polystyrene

	Consumption 1000 metric tons	
	1979	**1980**
Molding	864	734
Extrusion	598	507
Expandable bead	202	189
Export	73	84
Other (resold, blending)	80	94
Total	1,817	1,608
Rigid packaging	91	80
Housewares other than personal care	75	64
Toys	51	45
Radio/TV/stereo cabinets	38	41
Casettes	45	41
Disposable flatware, cutlery	41	41
Disposable tumblers, cocktail glasses	40	40
Building and construction	45	34
Furniture	42	30
Photographic	24	23
Personal care housewares	25	20
Disposable dishes, cups, bowls	27	20
Other appliances/electronics	31	18
Medical	18	18
Novelties	17	16
Small appliances	17	14
Blow molded packaging and disposables	15	14
Other recreational	19	14
Other consumer/industrial	21	14
Other furniture/furnishings	11	11
Refrigerators and freezers	13	10
Air conditioners	12	9
Toilet seats	10	9
Footwear (heels)	11	9
Produce baskets	9	7
Other injection molding	116	92
Total molding	864	734

(Continued)

Table 1.6. Uses of Polystyrene (Continued)

	Consumption 1000 metric tons	
	1979	**1980**
Vending and portion cups	84	73
Oriented film and sheet	85	64
Foam sheet stock food trays	57	56
Dairy containers	54	50
Lids	46	45
Foam sheet egg cartons	29	30
Foam board	31	27
Foam sheet single-service plates	25	25
Refrigerators and freezers	23	18
Other extrusion, solid	36	18
Foam sheet hinged containers (single service)	18	18
Plates and bowls, disposable	17	15
Building and construction, solid extrusion	18	11
Toys and recreational	11	9
Housewares	14	9
Miscellaneous consumer/industrial	10	9
Foam sheet single-service cups	9	9
Other foam sheet	13	9
Furniture and furnishings	9	7
Other appliances/electronics	9	5
Total extrusion	598	507
Cups and containers	63	60
Billets for building and construction	59	50
Shapes for packaging	39	39
Other shapes	18	17
Loose fill for packaging	14	14
Other billets	9	9
Total expandable bead	202	189

(Concluded)

Acrylic polymers are produced by the polymerization of acrylates such as methyl methacrylate, ethyl acrylate, and acrylonitrile. Polymethyl methacrylate offers light transmittance, color stability, toughness, dimensional stability, abrasion resistance, and weather stability. Applications include outdoor signs, interior and exterior lighting, automotive lenses, aircraft canopies, and architectural panels. Acrylic fibers are a prominent class of synthetic fibers.

Total sales of acrylics were 235,000 metric tons in 1980 (Table 1.8).

Table 1.7. Uses of Acrylonitrile-Butadiene-Styrene

	Consumption 1000 metric tons	
	1979	**1980**
Building and consruction	150	110
Transportation	100	68
Appliances	73	60
Business machines, telephones	30	26
Recreational	27	18
Consumer electronics	18	15
Luggage	11	10
Modifiers	11	9
Furniture	6	4
Packaging	4	2
Export	51	59
Other	82	60
Total	563	441

Table 1.8. Uses of Acrylic

	Consumption 1000 metric tons	
	1979	**1980**
Cast sheet	80	73
Molding and extrusion compound	71	61
Coatings	49	45
Other grades	39	31
Other (emulsion polymers, etc.)	29	25
Total	268	235
Construction	37	34
Industrial	19	17
Transportation	9	8
Signs	8	7
Consumer	7	7
Total cast sheet	80	73
Construction	29	28
Transportation	25	18
Industrial	14	13
Consumer	2	1
Signs	1	1
Total molding and extrusion	71	61

SECTION 1.5. ENGINEERING THERMOPLASTICS

Engineering thermoplastics are synthetic polymers which provide an improved property balance for a variety of stringent applications. These materials are generally characterized by improved mechanical properties, high temperature capability, and flame resistance, and offer improvements in processing flexibility, freedom of design, and overall assembly economics over metal, ceramics, and glass.

Polycarbonates are polyesters of carbonic acid, and are prepared by three major processes: transesterification, phosgenation of aromatic dihydroxy compounds, and interfacial polycondensation. They offer high temperature capability, high impact strength, high stiffness, good dimensional stability, high creep resistance, good electrical properties, low water absorption, high transparency, and good stain resistance. Applications include electric knife handles, housewares, appliance housings, electrical components, food vending machines, glazing, outdoor lighting, automotive parts, returnable milk containers, blood processing equipment, tractor cab parts, mine lighting guards, and business machine housings.

Total sales of polycarbonates were 99,000 metric tons in 1980 (Table 1.9).

Table 1.9. Uses of Polycarbonate

	Consumption 1000 metric tons	
	1979	**1980**
Glazing	32	31
Communications and electronics	22	21
Appliances	15	13
Sports and recreation	7	7
Transportation	8	6
Lighting	5	4
Signs	4	3
Other	14	14
Total	107	99

Polyphenylene oxide polymers are prepared by oxidative coupling of phenols such as 2,6-xylenol, and exhibit good thermal, mechanical, and electrical properties and chemical resistance. Modified polyphenylene oxide is the principal commercial material. Applications include high frequency insulation, appliance parts, electronic components, food packages, hospital utensils, pump components, automotive grilles, and plumbing fixtures.

Total sales of modified polyphenylene oxide were 53,000 metric tons in 1980 (Table 1.10).

Table 1.10. Uses of Modified Polyphenylene Oxide

	Consumption 1000 metric tons	
	1979	1980
Appliances	15	14
Business machines	13	13
Electrical/electronics	12	10
Transportation	12	9
Plumbing and hardware	3	3
Other	5	4
Total	60	53

Polysulfone polymers consist of phenylene units linked by three different chemical groups: isopropylidene, ether, and sulfone. Polysulfone contains aliphatic isopropylidene linkages; polyethersulfone and polyphenylsulfone are wholly aromatic. They are characterized by thermal stability, oxidation resistance, chemical resistance, and rigidity. Applications include electrical and electronic parts, aircraft interiors, battery cases, and food processing equipment.

Acetals are produced by polymerization of aldehydes, and exhibit strength, stiffness, creep resistance, toughness, chemical resistance, and electrical properties. Acetal homopolymers, such as polyformaldehyde, are used in business machine cams, gears, and printing wheels, automotive parts, ball cocks and shower heads in plumbing, furniture casters, bra buckles and clothing fitments, electric clocks, and lawn mower wheels. Acetal copolymers, such as those based on trioxane, are used in gears, cams, housings, fender extensions, valves and valve stems, tub assemblies, and toys.

Total sales of acetals were 43,000 metric tons in 1980 (Table 1.11).

The polyamides or nylons are prepared by polymerization of amino acids, or condensation of diamines with dibasic acids. The five major nylon homopolymers are:

nylon 6 (polycaprolactam), from caprolactam
nylon 6/6, from hexamethylenediamine and adipic acid
nylon 6/10, from hexamethylenediamine and sebacic acid
nylon 11, from 11-amino-undecanoic acid
nylon 12, from 12-amino-dodecanoic acid

These materials are characterized by outstanding resistance to repeated impact and fatigue, low friction coefficient, excellent abrasion resistance, and adequate electrical properties. Applications include packaging of foods, sausage casings, cable jacketing, braided tubing for automotive applications, appliance parts, floats, cams and gears, and fibers.

Total sales of nylon were 125,000 metric tons in 1980 (Table 1.12).

Table 1.11. Uses of Acetals

	Consumption 1000 metric tons	
	1979	**1980**
Consumer Products	10	10
Transportation	12	9
Industrial	9	8
Plumbing and hardware	8	8
Appliances	4	3
Electrical/electronics	2	2
Sheet, rod, and tube	2	2
Other	1	1
Total	48	43

Polyimides are produced by the reaction of an aromatic tetracarboxylic acid dianhydride with an aromatic diamine. They offer superior performance at higher temperatures and find their greatest use in high-temperature applications such as adhesives and glass-reinforced composites.

Polyetherimides are polymers which contain regular repeating ether and imide linkages. The repeating aromatic imide units are connected by aromatic ether units. They are characterized by high temperature capability, high mechanical strength, and good electrical properties which are stable across a wide range of frequencies and temperatures. They are designed for engineering applications in the transportation and construction industries.

Table 1.12. Uses of Nylon

	Consumption 1000 metric tons	
	1979	**1980**
Injection molding		
Transportation	34	28
Consumer products	18	17
Electrical/electronics	15	13
Industrial	15	13
Appliances	8	6
Extrusion		
Filaments	11	10
Film	9	10
Sheet, rod, and tube	6	5
Wire and cable	4	4
Export	13	10
Other	11	9
Total	144	125

Polybenzimidazoles (PBI) are produced by reactions of compounds such as 3,3′-diaminobenzidine and diphenylisophthalate or isophthalamide. They are designed for high-temperature applications such as glass-reinforced composites and adhesives.

Thermoplastic polyesters are produced by the polyesterification reaction between a glycol and a dibasic acid. Polyethylene terphthalate (PET) is prepared by reaction of either terephthalic acid or dimethyl terephthalate with ethylene glycol. Polybutylene terephthalate (PBT) is obtained from the condensation reaction of dimethyl terephthalate and 1,4-butanediol. Only part of the thermoplastic polyester produced is in the form of engineering grades. They are characterized by high strength and high temperature resistance.

Total sales of thermoplastic polyesters were 481,000 metric tons in 1980, of which 23,000 metric tons were engineering grades (Table 1.13).

Poly(amide-imide) is the condensation polymer of trimellitic anhydride and various aromatic diamines, and is characterized by high strength and good impact resistance. Applications include high-performance electrical/electronic parts, textile equipment, pumps, valves, and turbines.

Table 1.13. Uses of Thermoplastic Polyester

	Consumption 1000 metric tons	
	1979	1980
Nonengineering grades	382	458
Engineering grades	27	23
Total	409	481
Soft-drink bottles	150	200
Merchant film	110	119
Captive film	80	86
Coating for ovenable board	20	25
Sheeting for blisters	10	12
Strapping	5	6
Injection molding	5	5
Other blow-molded bottles	2	5
Total nonengineering grades	382	458
Electrical/electronics	9	9
Transportation	10	7
Plumbing and hardware	3	3
Appliances	2	2
Other	3	2
Total engineering grades	27	23

SECTION 1.6. OTHER THERMOPLASTICS

Cellulosic polymers are produced by chemical modification of cellulose. Cellulose nitrate is prepared by direct nitration of cellulose, and offers dimensional stability, low water absorption, and toughness. Unfortunately, it is flammable and lacks stability to heat and sunlight. Applications include personal accessories, toilet articles, and various industrial items.

Cellulose acetate is produced by reaction of cellulose with acetic acid and acetic anhydride. To produce cellulose acetate butyrate and cellulose acetate propionate, the acetic anhydride is partly replaced with butyric anhydride and propionic anhydride, respectively. Cellulose esters offer the advantageous combination of being both tough and transparent, and are used in film, sheet, and packaging applications.

Ethyl cellulose is produced by reaction of cellulose with caustic to form alkali cellulose, which then reacts with ethyl chloride to form ethyl cellulose. It offers superior toughness at low temperatures. Applications include football helmets, tool handles, electric appliance parts, and flashlight cases.

Total sales of cellulosics were 51,000 metric tons in 1980 (Table 1.14).

Acetals, nylons, and **thermoplastic polyesters** are commercially produced in nonengineering grades.

Table 1.14. Uses of Cellulosics

	Consumption 1000 metric tons	
	1979	**1980**
Molding and extrusion	29	24
Nonphotographic film and sheet	21	16
Extruded sheet for industrial use	15	11
Total	65	51

SECTION 1.7. THERMOSETTING POLYMERS

Polyurethanes are produced by the reaction of polyisocyanates with compounds containing reactive hydrogen atoms, and have found commercial use in flexible, semi-flexible, and rigid foams, in coatings, elastomers, and adhesives, and in spandex fibers.

Flexible polyurethane foams are used in furniture, bedding, and transportation seating. Total consumption was 510,000 metric tons in 1980 (Table 1.15). Rigid polyurethane foams are used in construction, appliances, and transportation, primarily as thermal insulation. Total consumption was 234,000 metric tons in 1980 (Table 1.15).

Polyurethane coatings offer toughness, hardness, mar-resistance, flexibility, chemical resistance, and resistance to abrasion. Polyurethane elastomers are used in textures for specialty upholstering and luggage, automotive fuel tanks, collapsible fuel storage cells, and wire and cable jacketing. Polyurethane adhesives are used in shoe construction.

Phenolic resins are prepared by the condensation of a phenol and an aldehyde, phenol-formaldehyde being the most common combination. They offer low cost and satisfactory all-around properties. Applications include appliance parts, telephone handsets, washing machine agitators, pump impellers, switches, oven parts, resistor casings, particle board, buttons, poker chips, coatings, adhesives, and plywood bonding.

Total sales of phenolic resins were 701,000 metric tons in 1980 (Table 1.16).

Table 1.15. Uses of Polyurethane Foams

	Consumption 1000 metric tons	
	1979	1980
Flexible foam	600	510
Rigid foam	256	234
Total	856	744
Furniture	230	200
Transportation	170	130
Bedding	85	80
Carpet underlay	60	50
Other	55	50
Total flexible foam	600	510
Building insulation	133	120
Household and commercial refrigeration	48	48
Industrial insulation	21	22
Transportation	22	18
Furniture	12	8
Other	20	18
Total rigid foam	256	234

Table 1.16. Uses of Phenolic

	Consumption 1000 metric tons	
	1979	1980
Plywood	232	210
Insulation materials	131	128
Molding compounds	141	120
Fibrous and granulated wood	54	45
Foundry and shell moldings	51	41
Building laminates	24	20
Coated and bonded abrasives	18	17
Friction materials	18	15
Furniture laminates	11	10
Protective coatings	10	9
Electrical/electronics laminates	9	9
Other laminates	6	5
Export	13	12
Other	70	60
Total	788	701

Amino resins are prepared by the reaction of amines, or compounds bearing amino groups, with aldehydes. Urea-formaldehyde and melamine-formaldehyde resins are the most commonly used. They offer moisture and solvent resistance, electrical properties, heat resistance, and mar resistance. Applications include electrical wiring devices, dinnerware, buttons, industrial laminates, and coating.

Total sales of amino resins were 589,000 metric tons in 1980 (Table 1.17).

Table 1.17. Uses of Urea and Melamine

	Consumption 1000 metric tons	
	1979	**1980**
Fibrous and granulated wood	463	402
Molding compounds	48	35
Textile treatment and coating resins	43	34
Plywood	39	33
Paper treatment and coating resins	35	30
Protective coatings	40	28
Laminating	13	12
Export	11	8
Other	8	7
Total	700	589

Epoxy resins are divided into three types: conventional epoxy, prepared by reacing epichlorohydrin with a polyhydroxy compound such as bisphenol A; epoxidized novolacs, such as epoxy cresol novolac and epoxy phenol novolac, prepared in a similar manner; and cycloaliphatic epoxy, prepared by the peracetic acid epoxidation of cyclic olefins. Epoxy resins offer toughness, high bond strength, and resistance to moisture and chemicals. Applications include adhesives, coatings, waterproofing membranes, and pipe fittings.

Total sales of epoxy resins were 144,000 metric tons in 1980 (Table 1.18).

Polyesters are produced by the reaction of dihydric alcohols (glycols) and dicarboxylic acids. Polyesters are classified as saturated or unsaturated, depending on the presence or absence of reactive double bonds in the linear polymer. Saturated polyesters, such as ethylene glycol terephthalate, find their greatest use in the production of fibers and film.

Table 1.18. Uses of Epoxy

	Consumption 1000 metric tons	
	1979	**1980**
Protective coatings		
Can and drum coatings	17	16
Plant maintenance	13	12
Auto primers	11	9
Pipe coatings	4	4
Appliance finishes	4	3
Other	18	16
Reinforced uses		
Electrical laminates	11	11
Filament winding	10	11
Other	6	6
Tooling, casting, and molding	14	12
Export	16	17
Other	9	12
Total	150	144

Unsaturated polyesters are used principally with fibrous reinforcement for fabricating a wide variety of articles, fibrous glass being the reinforcement most generally used. By design of the polyester molecule, properties such as fire retardance, corrosion resistance, and electrical properties can be achieved. Applications include roof panels, skylights, awnings, wall sidings, boat hulls, aircraft components, buttons, truck bodies, advertising displays, and structural lay-ups.

Total sales of polyesters were 440,000 metric tons in 1980, 340,000 metric tons of which were used for reinforced polyesters.

Alkyds are a class of polyesters which were developed primarily for high performance electronics and electrical systems. Alkyd molding compounds are combinations of liquid polyester resins (mixtures of unsaturated polyester and monomer) and various fillers.

Total sales of alkyds were 210,000 metric tons in 1980.

Allyl resins are esters of organic acids, such as the diallyl esters of phthalic, isophthalic, maleic, and chlorendic acids. They offer high temperature stability, chemical and moisture resistance, and good molding performance. Applications include electrical and electronic parts, impregnated glass cloth, decorative laminates, coatings, and sealants.

Silicone polymers are polymers with a backbone of alternating atoms of silicon and oxygen, with organic groups attached to the silicon atoms. They are produced by partial hydrolysis of methylchlorosilanes and subsequent stabilization. They offer thermal stability, surface properties, electrical properties, and inertness, Applications include mold release agents, defoamers, heat transfer fluids, electrical and electronic components, paper coatings, surfactants, curtain wall sealants, water repellents for masonry, and wire insulation.

Furan resins are prepared by the condensation polymerization of furfuryl alcohol. They offer chemical resistance and high char yield. Applications include chemical-resistant coatings, mortars, and grouts, adhesives, and laminates.

CHAPTER 2

Decomposition, Combustion, and Propagation

The entire process of thermal decomposition, combustion, and fire propagation is perhaps best described by considering it on three scales: micro, macro, and mass. The process can be described on the micro scale by discussing the behavior of the polymer molecule; on the macro scale by discussing the behavior of a unit mass of material, such as one gram; and on the mass scale by discussing the behavior of a complete system such as a room or structure.

SECTION 2.1. THE BURNING PROCESS ON A MICRO SCALE

The behavior of the polymer molecule with increasing temperature can be divided into five stages:

Stage I. **Heating.** Heat from an external source is applied to the polymer, progressively raising its temperature. There is relatively little change in physical properties.

Stage II. **Transition.** Within a relatively narrow temperature range, called the glass transition temperature, the polymer changes from a relatively hard and brittle material to a viscous or rubbery condition. The mechanical properties, and some thermal properties, change rapidly above this temperature. In general, the load-bearing capabilities of the material decrease above this temperature (Table 2.1).

Stage III. **Degradation.** The polymer chain is no sronger than its weakest link, and the temperature of initial degradation is usually the

Table 2.1. Glass Transition Temperatures for Various Polymers

	Tg, °C
Polyethylene	−125
Polypropylene	−20
Polybutylene	−25
Polybutadiene	−85
Poly(4-methyl-1-pentene)	29
Polytetrafluoroethylene	−113,127
Polychlorotrifluoroethylene	45
Polyvinyl chloride	80
Polyvinyl fluoride	−20
Polyvinylidene chloride	−18
Polystyrene	100
Polymethyl methacrylate	50
Polyformaldehyde	−85
Polyacetaldehyde	−30
Polycaprolactam (nylon 6)	75
Polyhexamethylene adipamide (nylon 6/6)	57
Polyhexamethylene sebacamide (nylon 6/10)	50
Polypropylene oxide	−75
Polycarbonate of bisphenol A	149
Poly(dimethyl siloxane)	−123
Polysulfone	190
Polyethersulfone	221
Polyethylene terephthalate	70
Polybutylene terephthalate	40

temperature at which the least thermally stable bonds fail. The bulk of the polymer may be stable, but the failure of the weakest bonds often produces results such as discoloration. Degradation may be of two types: thermal degradation, in the absence of oxygen, and thermal-oxidative degradation, influenced by both heat and oxygen.

The characteristics of the polymer which are important in this stage are:

1. Decomposition temperature of the least thermally stable bonds.
2. Proportion of the least stable bonds in the polymer.
3. Latent heat of decomposition of the least stable bonds.

Endothermic decomposition absorbs heat and slows the temperature rise, while exothermic decomposition supplies heat and accelerates the temperature rise.

Stage IV. **Decomposition.** The majority of the bonds existing in the polymer reach the failure point, and the polymer mass itself changes. The polymer may exhibit behavior ranging from disintegration, with complete loss of physical integrity, to rearrangement with little weight loss to give different properties. An example of the former is the complete depolymerization of some polymers, such as polyformaldehyde and polymethyl methacrylate. An example of the latter is the thermal modification of polyacrylonitrile to yield "black orlon."

The degradation and decomposition stages can be separated only when the failure temperature of the least stable bonds is significantly lower than the decomposition temperature of the majority of the bonds in the polymer. When a polymer contains a variety of bonds with an almost continuous spectrum of decomposition temperatures, the degradation and decomposition stages merge into a single stage.

The decomposition products vary with polymer composition, temperature level, rate of temperature rise, endotherms and exotherms, and rate of volatiles evolution. Differential thermal analysis, thermogravimetric analysis, and time-of-flight mass spectrometer analysis indicate that relatively small differences in chemical composition, temperature, and heating rate can drastically change the nature of the decomposition products. It is obvious that actual fires involve varying and changing temperatures well above the normal decomposition temperature of many polymers; the products obtained in careful laboratory decompositions will therefore not necessarily be the same as those produced at actual fire temperatures. Even laboratory studies cover a range of decomposition temperatures, rather than a specific level considered to be the decomposition temperature (Table 2.2).

The decomposition of the polymer produces two types of materials: the polymer chain residue, which continues to provide some structural integrity, if such a residue is obtained, and polymer fragments which are highly vulnerable to oxidation. Access to oxygen to the heated carbonaceous residue can result in glowing of the char, but flaming combustion generally occurs in the gas phase near the polymer residue, and involves gaseous and finely divided solid material.

The decomposition stage is significantly affected by the following characteristics of the polymer:

1. Decomposition temperature of the various bonds comprising the bulk of the polymer.

2. Latent heat of decomposition of the various bonds. If the decomposition is on the average endothermic, heat must be supplied from

Table 2.2. Decomposition Temperatures for Various Polymers

	Decomposition Temperature Range, °C
Polyethylene (PE)	335–450
Polypropylene (PP)	328–410
Polyisobutene	288–425
Polytetrafluoroethylene (PTFE)	508–538
Polychlorotrifluoroethylene (CTFE)	347–418
Polyvinyl acetate	213–325
Polyvinyl alcohol	250
Polyvinyl butyral	300–325
Polyvinyl chloride (PVC)	200–300
Polyvinyl fluoride (PVC)	372–480
Polyvinylidene chloride	225–275
Polyvinylidene fluoride	400–475
Polystyrene	285–440
Styrene-butadiene copolymer	327–430
Polymethyl methacrylate	170–300
Polyacrylonitrile	250–280
Cellulose triacetate	250–310
Polyformaldehyde	222
Polyethylene oxide	324–363
Polypropylene oxide	270–355
Nylon 6 and 6/6	310–380
Poly-p-xylylene	420–465
Polyethylene terephthalate	283–306
Polycarbonate	420–620

other sources to continue decomposition. If the decomposition is on the average exothermic, the decomposition can maintain itself.

3. Decomposition behavior, or the manner in which the polymer decomposes. This involves the relative amounts of combustible and non-combustible fragments.

Stage V. **Oxidation.** At a high enough temperature and in the presence of sufficient oxygen, the oxidation of the polymer fragments proceeds rapidly enough to produce heat and flame in the gas phase, and perhaps glowing in the solid residue.

SECTION 2.2. THE BURNING PROCESS ON A MACRO SCALE

Consideration of the burning process on the macro scale differs from that on the micro scale in that the basic polymer alone is considered on the micro scale, while the plastic material, including fillers or blowing agents, is considered on the macro scale.

The burning of an individual unit mass of material, such as one gram, can be considered as occurring in five stages:

Stage I. **Heating.** Heat from an external source is applied to the material, progressively raising its temperature. The external supply of heat may come from direct exposure to flame (radiation and convection) in the case of material at an exposed surface, from heat transfer from hot fire gases (conduction and convection), or from a hot solid mass (conduction) as in the case of an adhesive behind exposed material or a plastic covered by a protective coating or covering. The rate of temperature rise is a function of the flow rate of applied heat, of the temperature differential, and of the following characteristics of the material:

1. Specific heat, or the amount of heat required to raise the temperature of the unit mass by one unit of temperature measurement. Materials with high specific heats increase more slowly in temperature than low-specific-heat materials (Table 2.3).

2. Thermal conductivity, or the rate at which heat flows through a given thickness of material under a given temperature differential. A high thermal conductivity means that heat is transferred to an adjacent unit mass more rapidly than would be the case with a low thermal conductivity (Table 2.4).

3. Latent heat of fusion (melting), vaporization, or other changes which may occur in the material during the heating process.

Stage II. **Decomposition.** The material reaches its decomposition temperature, and begins to evolve one or more of the following types of products:

1. Combustible gases, or gases which will burn in the presence of air. This group would include methane, ethane, ethylene, formaldehyde, acetone, and carbon monoxide.

2. Noncombustible gases, or gases which normally would not burn in the presence of air. This group would include carbon dioxide, hydrogen chloride, and hydrogen bromide.

3. Liquids, usually partially decomposed polymer.

4. Solids, in the form of carbonaceous residue or char.

5. Entrained solid particles, or polymer fragments, which appear as smoke.

Table 2.3. Specific Heats for Various Materials

	Specific Heat cal/gm °C
Polyethylene (PE)	0.55
Ethylene-vinyl acetate copolymer (EVA)	0.55
Ethylene-ethyl acrylate copolymer (EEA)	0.55
Polypropylene (PP)	0.46
Polypropylene, rubber modified	0.5
Ethylene methacrylic acid ionomer	0.55
Polytetrafluoroethylene (PTFE)	0.25
Fluorinated ethylene-propylene (FEP)	0.28
Polychlorotrifluoroethylene (CTFE)	0.22
Polybutylene	0.45
Methylpentene polymer	0.52
Polyvinyl chloride (PVC)	0.20–0.28
Polyvinylidene chloride	0.32
Polyvinylidene fluoride	0.33
Polystyrene (PS)	0.32
Polystyrene, 20–30% glass filled	0.23–0.27
Styrene-acrylonitrile copolymers (SAN)	0.32–0.34
Styrene-butadiene thermoplastic elastomers	0.45–0.50
Acrylonitrile-butadiene-styrene (ABS)	0.30–0.40
Polymethyl methacrylate	0.35
Cellulose nitrate	0.30–0.40
Cellulose acetate	0.30–0.50
Cellulose acetate butyrate	0.30–0.40
Cellulose propionate	0.30–0.40
Ethyl cellulose	0.30–0.75
Acetal	0.35
Nylon 6	0.38
Nylon 6/6	0.40
Nylon 6/10	0.40
Nylon 6, 20–40% glass filled	0.30–0.35
Nylon 11	0.58
Nylon 11, glass filled	0.42
Polycarbonate	0.30
Phenoxy	0.40
Polysulfone	0.31
Polyphenylene oxide, modified	0.32
Polyimide, aromatic	0.27
Phenol-formaldehyde	0.38–0.42
Phenolic, asbestos filler	0.30

(Continued)

Table 2.3. Specific Heats for Various Materials (Continued)

	Specific Heat cal/gm °C
Melamine-formaldehyde, cellulose filler	0.40
Urea-formaldehyde, cellulose filler	0.40
Epoxy	0.25
Epoxy, silica filled	0.20–0.27
Polyester, chopped glass filler	0.25
Polyurethane	0.40–0.45
Allyl	0.26–0.55
Silicone, glass fiber filled	0.24–0.30

(Concluded)

Since in most cases ignition and combustion occur in the gas phase, it is obvious that complete elimination of combustible gases would effectively preclude burning. In most cases, unfortunately, this is impossible, since most organic materials cannot be reduced to a highly carbonaceous residue without releasing some volatile comounds containing hydrogen.

Noncombustible gases are of course preferred to combustible gases from the viewpoint of burning. The evolution of any gases, however, results in expansion of the system, into adjacent unit masses or into the adjacent atmosphere, or both, and by disrupting the chemical and physical structure of the material can increase the degree of exposure to destructive temperatures. Some gases also present the additional disadvantage of irritation or toxicity.

Liquids may not be as combustible as gases, but they present possible hazards in spreading the unit mass so that it affects unit masses other than those originally adjacent to it, and have the potential of being converted into gases through a temperature rise and change in state.

The solid residue is the most desirable product of decomposition, because it helps preserve structural integrity and protect adjacent unit masses from decomposition, and impedes the mixing of air with combustible gases. The extent of weight loss on decomposition is therefore rightly used as one indication of resistance to fire.

Decomposition does physical violence to the polymer in varying degrees. Fragments of the polymer can be broken off from the mass and entrained in moving gases. These fragments appear as smoke particles, and, if swept into the flame itself, as incandescence.

Table 2.4. Thermal Conductivity of Various Materials

	Thermal Conductivity 10^{-4} cal./sec.–cm.^{2} (°C./cm.)
Polyethylene, low density	8.9
Polyethylene, medium density	8.0 to 10.0
Polyethylene, high density	8.0 to 12.4
Polypropylene	2.8
Polypropylene, rubber modified	3.0 to 4.0
Ethylene-methacrylic acid ionomer	5.8
Methylpenetene polymer	4.0
Polytetrafluoroethylene (PTFE)	6.0
Fluorinated ethylene-propylene (FEP)	5.9 to 6.0
Polychlorotrifluoroethylene (CTFE)	4.7 to 6.0
Polyvinyl chloride (PVC)	3.0 to 7.0
Polyvinylidene chloride	3.0
Polyvinylidene fluoride	3.0
Polystyrene	1.9 to 3.3
Styrene-acrylonitrile copolymers (SAN)	2.9 to 3.0
Styrene-butadiene thermoplastic elastomers	3.6
Acrylonitrile-butadiene-styrene (ABS)	4.5 to 8.0
Polymethyl methacrylate	4.0 to 6.0
Cellulose nitrate	5.5
Cellulose acetate	4.0 to 8.0
Cellulose acetate butyrate	4.0 to 8.0
Cellulose propionate	4.0 to 8.0
Ethyl cellulose	3.8 to 7.0
Acetal homopolymer	1.6 to 5.5
Acetal copolymer	5.5
Nylon 6	5.9
Nylon 6/6	5.8
Nylon 6/10	5.2
Nylon 6, glass-filled, 20–40%	5.2
Nylon 11	7.0
Nylon 11, glass filled	8.8
Polycarbonate	4.6
Polycarbonate, glass filled, 10–40%	2.5 to 5.2
Phenoxy	4.2
Polyphenylene oxide	4.5
Polyphenylene oxide, 30% glass filled	3.4
Polyphenylene oxide, modified	5.2
Polyphenylene oxide, modified, 20–30% glass filled	3.8
Chlorinated polyether	3.1

(Continued)

Table 2.4. Thermal Conductivity of Various Materials (Continued)

	Thermal Conductivity 10^{-4} cal./sec–cm.^{2} (°C./cm.)
Phenol-formaldehyde	3.0 to 6.0
Phenolic, asbestos filled	8.4
Melamine-formaldehyde, asbestos filled	13 to 17
Melamine-formaldehyde, cellulose filler	7 to 10
Urea-formaldehyde, cellulose filler	7 to 10
Epoxy	4 to 5
Epoxy, silica filled	10 to 20
Polyester	4
Polyester, glass fiber filled	10 to 16
Alkyd, glass filled	15 to 25
Polyurethane	1.5 to 7.4
Allyl	4.8 to 5.0
Allyl, glass filled	5 to 15
Silicone	3.5 to 7.5
Silicone, glass filled	7.5

(Concluded)

The decomposition state is significantly affected by the following characteristics of the material:

1. Temperature of initial decomposition, or the lowest temperature at which decomposition occurs. If the temperature of the unit mass never reaches this level, decomposition does not occur and the burning process is terminated before ignition even occurs. Where two polymers of equal specific heat and equal thermal conductivity are exposed to heat at a surface, the extent to which decomposition proceeds into the mass is largely a function of initial decomposition temperature.

2. Latent heat of decomposition, or the heat absorbed or released during decomposition. If decomposition is exothermic, the heat released increases the rate of temperature rise. If decomposition is endothermic, the heat absorbed is removed from the available supply, slowing the heating process.

3. Decomposition behavior, or the manner in which the polymer decomposes. This involves the relative amounts of combustible and noncombustible gases, liquids, solid residue, and solid particles, as well as the sequence of phase changes.

Stage III. **Ignition.** The combustible gases ignite in the presence of sufficient oxygen or oxidizing agent, and combustion begins. Ignition depends on the presence of an external source of ignition, such as flames or sparks, and the temperature and composition of the combined gas phase. The ignition stage is significantly affected by the following characteristics of the material:

1. Flash ignition temperature, or the temperature at which the gases evolved from the material can be ignited by a spark or flame. This is generally higher than the initial decomposition temperature (otherwise there would be no gases to ignite). (Table 2.5).

2. Self-ignition or auto-ignition temperature, or the temperature at which reactions within the material become self-sustaining to the point of ignition. This is generally higher than the flash-ignition temperature, because more energy is required to initiate self-sustaining decomposition than externally-sustained decomposition. (Table 2.5). One noteable exception is cellulose nitrate.

3. Limiting oxygen concentration, or the minimum level of oxygen required to sustain ignition and combustion. For normal use, a material can be considered self-extinguishing if it cannot continue burning with less than 21 percent oxygen, and non-igniting if it cannot be ignited with less than 21 percent oxygen (Table 2.6). The conditions under which a material appears "non-igniting" or "self-extinguishing," however, vary with the severity of the exposure.

Stage IV. **Combustion.** Combustion of the unit mass generates a certain amount of combustion heat (Table 2.7). The heat of combustion raises the temperature of the gaseous products of combustion and of the noncombustible gases, increasing heat transfer by conduction; expansion of the heated gases increases heat transfer by convection. Heating the entrained solid particles to incandescence increases heat transfer by radiation, and heating the solid residue increases heat transfer by conduction.

On the small scale of the unit mass, this stage represents full-scale or fully developed burning. Once the burning process has reached this stage, extinguishment, rather than inhibition, becomes critical. The most important characteristic of the material in this stage is its heat of combustion, the amount of heat released by the combustion of a unit mass. The net heat of combustion is the heat released by the combustion reactions, reduced by the amount of heat required to bring the unit mass from its initial state to the combustion stage. If the net heat of combustion is negative, then an external supply of heat is required to continue combus-

Table 2.5. Ignition Temperatures of Various Materials

Material	**Flash-Ignition Temp.**		**Self-Ignition Temp.**	
	°C.	**°F.**	**°C.**	**°F.**
Cotton	230–266	446–511	254	490
Paper, newsprint	230	445	230	445
White pine, shavings	228–264	406–507	260	500
Long leaf pine	220–230	428–446		
Red oak			416	781
Douglas fir	260	500		
Wool	200	401		
Polyethylene	341–357	645–675	349	660
Polypropylene, fiber			570	1058
Polytetrafluoroethylene			530	986
Polyvinyl chloride	391	735	454	850
Polyvinyl chloride-acetate	320–340	608–644	435–557	815–1035
Polyvinylidene chloride	532	990+	532	990+
Polystyrene	345–360	653–680	488–496	910–925
Polystyrene beads	296	565	491	915
Polystyrene, foam beadboard	346	655	491	915
Acrylonitrile-butadiene-styrene (ABS)			466	781
Styrene-acrylonitrile copolymer	366	690	454	850
Styrene-methyl methacrylate copolymer	329	625	485	905
Polymethyl methacrylate	280–300	536–752	450–462	832–864
Acrylic, fiber			560	1040
Cellulose nitrate	141	286	141	286
Cellulose acetate	305	581	475	887
Cellulose triacetate, fiber			540	1004
Ethyl cellulose	291	555	296	565
Polyamide (nylon)	421	790	424	795
Nylon 66, fiber			532	990
Polycarbonate	467	873	580	1076
Phenolic, glass fiber laminate	520–540	986–1004	571–580	1060–1076
Melamine, glass fiber laminate	475–500	887–932	623–645	1153–1193
Polyester, glass fiber laminate	346–399	655–750	483–488	811–910
Polyurethane, polyether, rigid foam	310	590	416	780
Silicone, glass fiber laminate	490–527	914–981	550–564	1022–1047

Table 2.6. Oxygen Index for Various Materials

		Thickness in.	Oxygen Index
Polyethylene,	Allied 1220		17.4
	Phillips Marlex 5002		17.5
	Phillips Marlex 5002, 50% Norton 38900		19.6
	Phillips Marlex 5002, 60% Alcoa C-74		30.2
	Du Pont Rulan 2,3,4		22
	Eastman Tenite 860	0.125	18
	ICI Alkathene WJG 11		18.4
	ICI Rigidex 2000		18.6
	Raychem Flamolin 419,851		24
	Raychem Flamolin 853		25
	Raychem Flamolin 511,711		26
	Raychem Flamolin 855		27
	Raychem Flamolin 610		28
	Raychem Flamolin 752, 511 GR		29
	coating grade		18.0
	unidentified, U.S.		17.4
	unidentified, France		17.3 to 23.9
Polypropylene	Hercules Profax 6505		17.4
	Hercules Profax P 7401		17.8
	Hercules Profax 6 X 23, 6 X 24	0.125	18
	Hercules Profax PC 072	0.125	18
	Hercules Profax SA 595	0.125	27
	Phillips Marlex HRV-120-01	0.125	27
	Phillips Marlex HGR-120-01	0.125	28
	Allied Plaskon 1050		25.6
	Allied Plaskon 1060		26.4
	Allied Plaskon 1070X		23.7
	Allied Plaskon FR-1050	0.125	24
		0.25	25
	Allied Plaskon FR-1052	0.125	26
	Allied Plaskon FR-1060	0.125	25
	Allied Plaskon FR-1080	0.125	28
	Avisun 2056		23.8
	Avisun 2356		25.2, 29.2
	Eastman Tenite 4221	0.125	17
	Eastman Tenite 423R	0.125	24
	Enjay Escon 185		24.5

(Continued)

Table 2.6. Oxygen Index for Various Materials (Continued)

		Thickness in.	Oxygen Index
Polypropylene,	Enjay Exxon D-569, D-570	0.125	27
	Enjay Exxon E-185, E-187	0.125	25
	Enjay Exxon CD495, CD495A	0.125	28
	ICI Propathene GSE 108		18.6
	ICI Propathene PXC 4515		25.8
	Rexall PP 11-R5		24.9
	LNP MF 1006	0.125	18
	LNP MF 1004 FR	0.125	27
	Raychem Flamolin 226		24
	Raychem Flamolin 227		26
	Raychem Flamolin 227 GR		28
	Union Carbide JMDE 9490		24.9
	Union Carbide JMDC 4400		20.5
	Union Carbide JMDA 9490		28.2
	unidentified, 30% glass		18.5
	unidentified, France		17.4 to 24.6
	fabric, 6.5 oz/sq. yd.		18.6
Polybutadiene, crosslinked			18.3
Polyallomer, Eastman Tenite PA 5021		0.125	17
Chlorinated Polyethylene			21.1
PTFE, Allied Halon G80, G700		0.125	95
Du Pont Teflon 6,6A,6C,6H,7A,7C,8,9		0.125	95
CTFE, Allied Plaskon 2200, 2300		0.125	95
LNP 803		0.125	83
Ethylene CTFE, Allied Halar		0.125	60
Ethylene TFE, Du Pont Tefzel 200		0.125	30
FEP, Du Pont Teflon 100, 110, 160		0.125	95
LNP 906		0.125	95
Polyvinyl chloride,	BFG Geon 101		45.0
	BFG Geon 8750	0.25	52
	BFG Geon Hi Temp 3005		50
	BFG Geon Hi Temp 3007	0.25	60
	BFG Geon Hi Temp 3010	0.25	63
	Plaskon X2013		31.5
	Plaskon 2005		40.3
	Union Carbide QSAH-7		40
	Sta-Flow B361	0.062	30
	Sta-Flow 2016, 2025, 5011	0.062	32
	Sta-Flow 7021, 2020	0.062	33
	Sta-Flow 2003A		36
	LNP VF-1003	0.125	42
	coating grade		26.0
	fabric, Phovyl 55, 6.5 oz/sq.yd		37.1

(Continued)

Table 2.6. Oxygen Index for Various Materials (Continued)

		Thickness in.	Oxygen Index
Polyvinyl chloride,	unidentified, U.S.		47.0
	unidentified, rigid, France		43.6 to 80.7
	unidentified, flexible, France		20.6 to 40.4
Polyvinyl alcohol, Du Pont Elvanol 70-05			22.5
Polyvinyl fluoride, Tedlar			22.6
Polyvinylidene chloride, Dow Saran 281 S 905			60.0
Polyvinylidene fluoride, Kynar			43.7
Polystyrene,	Monsanto Lustrex HT 95	0.125	18
	Monsanto Lustrex HT 88, LP 59-2	0.125	17
	Cosden 825 TV		17.7 to 19.2
	Richardson FR 650 8L		22.5
	Richardson FR GPS 201-L8		23.3
	Shell 325		17.8
Polystyrene,	Koppers 8		18.0
	Koppers Dylene 8X		19.0
	Koppers Dylene 85		25.2
	Koppers KDP 986-B		21.5
	Koppers KPD 986-E		23.5
	Westlake high-temperature		18.3
	unidentified, U.S.		18.1 to 18.3
	unidentified, France		18.0 to 21.1
ABS,	BFG Abson 89129		25.8 to 27.0
	BFG Abson 89140		18
	Durez 2311		25.2
	Marbon Cycolac T		18.8
	Marbon Cycolac T, TA, TB, GSE, EP 3510	0.125	19
	Marbon Cyclolac DH	0.125	20
	Marbon Cyclolac KJB	0.125	29
	Marbon Cyclolac KT		25.8
	Marbon Cyclolac KL		27.4
	Marbon Cyclolac KA		31.7
	Marbon Cycovin KA Natural, KA White	0.125	32
	Marbon Cycloloy 800	0.125	21
	Marbon Cycloloy KHP, KHS	0.125	30
	Marbon X-88		27.4
	Marbon X-91		33.5
	Monsanto Lustran 240, 440, 640, 643, 740	0.125	18
	LNP AF 1006	0.125	19

(Continued)

Table 2.6. Oxygen Index for Various Materials (Continued)

		Thickness in.	Oxygen Index
ABS,	Rexene 500 FR-1, 500 FR-2		28
	Uniroyal Royalite 20		18.3 to 19.7
	unidentified, 20% glass		21.6
	unidentified, France		18.1 to 39.0
SAN,	Monsanto Lustran SAN 21	0.125	18
	Fiberfil Acrylafil G40/30FR	0.125	24
	LNP BF 1004 FR	0.125	27
	LNP BF 1006 FR	0.125	28
	LNP BF 1006	0.125	19
	unidentified, U.S.		18.1 to 19.1
Acrylic,	Plexiglas		17.3
	Plexiglas G	0.030	16.7
		0.050	16.8
		0.125	17.4
	ICI Perspex		18.7
	ICI Perspex EM		20.2
	ICI Perspex FR		25.1
	ICI Diakon LG, LH, MG, MH	0.125	19
	unidentified, France		18.2 to 23.9
Acrylic,	fabric, Acrilan, 6.5 oz/sq.yd		18.2
	fabric, unidentified		19.6
Modacrylic,	fabric, Dynel, 6.5 oz/sq.yd.		26.7
	fabric, Dynel		27.3
	fabric, Dynel, glass filament		29.8
	fabric, Verel		29.8
Wool,	fabric, 7.0 oz/sq.yd		25.2
	fabric		23.8
Wood,	pine, France		22.4
	poplar, France		22.5
	oak, France		24.6
	fiber board, France		22.1
	particle board, France		24.5
Plywood,	France		25.4
Plywood,	FR, France		73.6
Cardboard			24.7
Cotton,	fabric, 6.5 oz/sq.yd		20.1
	fabric		18.6
	fabric, treated, US Army		27.3

(Continued)

Table 2.6. Oxygen Index for Various Materials (Continued)

	Thickness in.	Oxygen Index
Rayon, fabric, 6.5 oz/sq.yd.		19.7
fabric		18.9
felt		18.7
Cellulose acetate, Eastman Tenite 080	0.125	19
Eastman Tenite 091	0.125	27
0.1% water		16.8
4.9% water		18.1
fabric, 6.5 oz/sq.yd		18.6
Cellulose triacetate, fabric, Arnel, 6.5 oz/sq.yd		18.4
Cellulose butyrate, 0.06% water		18.8
2.8% water		19.9
Cellulose acetate butyrate, Eastman Tenite 265	0.125	20
Eastman Uvex	0.125	18
unidentified, U.S.		19.6
Cellulosics, unidentified, France		20.1 to 28.6
Acetal, Du Pont Delrin		14.7 to 16.1
Celanese Celcon		14.8 to 14.9
LNP KF 1006	0.125	16
unidentified, 30% glass		15.6
unidentified, U.S.		16.2
Nylon, unidentified, France		21.6 to 28.9
fabric, 6.5 oz/sq.yd		20.1
Nylon 6, Allied Plaskon 8230, HS-1	0.125	23
Allied Plaskon 8202C	0.125	28
AKZO Akulon K2-6V	0.125	22
AKZO Akulon M2-ZG340	0.125	26
AKZO Akulon K2-ZG340	0.125	28
BASF Ultramid B3K	0.125	25
Fosta 339,438,446,471,475,512,578,589, 851,981	0.125	25
LNP PF 1006 FR	0.125	28
Nylon 6-6, AKZO Akulon R600	0.125	29
BASF Ultramid A3K	0.125	25
BASF Ultramid A3XG5, A3XG7	0.125	30
Du Pont Zytel 103 series		27
Du Pont Zytel 101,102,105,106,121,122, 131,42,43 series		28
Du Pont Zytel 70G33LG, 71G33L		24

(Continued)

Table 2.6. Oxygen Index for Various Materials (Continued)

	Thickness in.	Oxygen Index
Nylon 6-6, Fiberfil Xylon 66N1, 6601	0.125	29
Fiberfil Nylafil G1-30	0.125	21
Fiberfil Nylafil G1-30-FR	0.125	38
Foster Grant Fosta 436	0.125	25
ICI Maranyl A190 Nat, A192 Nat	0.125	23
ICI Maranyl A100, A101, A105, A150, AD146, AD148	0.125	26
ICI Maranyl AD197	0.125	28
Monsanto Vydyne M340, R100, R200	0.125	21
Monsanto Vydyne 10V, 10X, 11V, 20M, 20N, 20V, 20X, 21, 21X	0.125	25
LNP RF 1004	0.125	21
LNP RF 100-1 to 6-HS, RF 100-10	0.125	22
LNP RF 1006 FR	0.125	28
dry		24.3
8% water		30.1
unidentified, U.S.		28.7
Nylon 6-10, Monsanto Vydyne 50V	0.125	25
LNP QF 1006 FR	0.125	28
Nylon 6-12, Du Pont Zytel 151, 151L, 153-HS-L, 157-HS-LBK-10, 158, 158L		25
Aromatic polyamide, Nomex		26.7
Nomex N-4274, fabric, 4.8 oz/sq.yd		28.2
Polyimide, Kapton 100H Polyimide, Kapton 100H		36.5
Polybenzimidazole, unidentified		40.6
Chlorinated polyether, Hercules Penton		23.2
Polyethylene oxide, Union Carbide Polyox WSR-35		15.0
Polycarbonate, GE Lexan 101,111,112,121,144	0.125	25
GE Lexan 141		24.9 to 27.0
GE Lexan 191		22.5
GE Lexan 2014		31.0 to 32.0
GE Lexan 2816		37
GE Lexan 3412	0.125	30
GE Lexan 3412-131, 20% glass		29.8
GE Lexan 3414	0.125	29
GE Lexan 3414-131, 40% glass		29.3
GE Lexan 500	0.125	33

(Continued)

Table 2.6. Oxygen Index for Various Materials (Continued)

		Thickness in.	Oxygen Index
Polycarbonate,	GE Lexan NB155	0.125	44
	GE Lexan DL444		39.7
	GE Lexan DL444, 20% glass		42.0
	GE Lexan 8000-111		21.3
	GE Lexan 8070-112		22.2
	GE Lexan sheet, glass		27.8
	GE Lexan commercial sheet		26.1
	Mobay Merlon M40, M50, M60	0.125	25
	Mobay Merlon SE-2200	0.125	30
	Mobay Merlon SE-3200	0.125	29
	Mobay Merlon 9310	0.125	31
	Fiberfil Carbasar J-54/10	0.125	33
	Fiberfil Polycarbafil G50/20, G50/30, G50/40	0.125	29
	Fiberfil Polycarbafil G50/20/FR	0.125	35
	LNP DF 1004	0.125	27
	LNP DF 1006	0.125	30
	LNP DF 1008	0.125	29
Polysulfone,	Union Carbide Udel P-1700, P-1710	0.080	30
	Union Carbide Udel P-1720	0.080	32
	Union Carbide experimental,		
	PSF/Sb/B		33
	Cl_2BIS/PSF		43
	Cl_2PSF		41
	Cl_4PSF		51
	Fiberfil Sulfil G1500/20	0.125	33
	LNP GF 1006	0.125	35
	LNP GF 1006 FR	0.125	39
	LNP GF 1008	0.125	36
Modified polyphenylene oxide,	GE PPO		29
	GE PPO 534		30.0
	GE GP-275		24
	GE Noryl SE-1		28.0 to 29.0
	GE Noryl SE-100		27.4 to 33.0
	GE Noryl 731		24.0 to 24.3
	GE Noryl SE1GNF2, SE1GFN3		27
	LNP ZF 1004 FR	0.125	28

(Continued)

Table 2.6. Oxygen Index for Various Materials (Continued)

		Thickness in.	Oxygen Index
Phenolic,	Allied Plaskon Phenall 8000	0.125	60
	Allied Plaskon Phenall 8020	0.125	45
	Allied Plaskon Phenall 8700	0.125	55
	Hooker Durez 18441	0.125	26
	Hooker Durez 791,792,22486	0.125	28
	Hooker Durez 16744	0.125	31
	Hooker Durez 16274	0.125	32
	Hooker Durez 1308,1328	0.125	34
	Hooker Durez 22829	0.125	36
	Hooker Durez 21426,22065	0.125	37
	Hooker Durez 22042	0.125	39
	Hooker Durez 22262	0.125	40
	Hooker Durez 19089	0.125	44
	Hooker Durez 11864	0.125	58
	Hooker Durez 18975	0.125	66
	GE Genal 4000SE	0.125	18
	GE Genal 4200SE	0.125	27
	GE Genal 14020,14029,4050,4200	0.125	29
	GE Genal 12553,12983,4300,4301	0.125	36
	GE Genal 12992	0.125	39
	Rogers RX-525	0.125	32
	Rogers RX-462	0.125	59
	Rogers RX-468	0.125	64
	Rogers RX-610N	0.125	40
	mica filled, Plenco 343-B817		52.8
	paper laminate, Textolite 11571B		21.7
Epoxy,	Epon 826/MNAA		20.8
	Epon 826/HHPA/DMP-30		19.8
	Epon 826/HHPA/DMP-30, 50% Norton 38900		25.0
	Epon 826/HHPA/DMP-30, 60% Alcoa C-333		40.8
	Plenco 200D-B1242		38.3
	Plenco 2000F-B6432		47.6
	Araldite 6005/MNA/DMP-30		21.7
	Araldite 6005/HHPA/DMP-30		20.3
	Araldite 6005/TETA		23.2
	Araldite 6005/TONOX		27.1
	Araldite 6005/BF3/MEA		23.8
	Araldite 6005/DMP-30, 6060/MNA/DMP-30		21.1

(Continued)

Table 2.6. Oxygen Index for Various Materials (Continued)

		Thickness in.	Oxygen Index
Epoxy,	CY-175/MNA/SO		18.3
	ERLA 4221/MNA/SO		19.8
	glass laminate, GE 1158G10		24.9
	glass laminate, GE 11673FR-2		32.6
	glass laminate, GE 11635FR-4		49.0
Polyester,	fabric, 6.5 oz/sq.yd, polyethylene terephthalate		20.6
	fabric, Dacron T-54, polyethylene terephthalate		21.0
	Hooker Hetron 92		30.4
	Hooker Hetron 92, 5% zinc borate		34.1
	Hooker Hetron 92, 5% antimony oxide		42.3
	Koppers 3465-5	0.125	28
	Koppers 3404-5	0.125	31
	Koppers 3404-15,3433-5,3463-5,3464-5	0.125	32
	Koppers 3401-5	0.125	33
	Koppers 3401-15	0.125	34
	Koppers 3401-25	0.125	36
	Koppers 3400-30	0.125	38
	Diamond Shamrock Dion 6692		32
	Diamond Shamrock Dion 6692, 6% zinc borate		33
	Diamond Shamrock Dion 6692, 6% antimony oxide		36.9
	Atlas Atlac A711-050		28.1
	Atlas Atlac A711-050, 5% zinc borate		32.9
	Atlas Atlac A711-050, 5% antimony oxide		37.3
	Atlas Atlac 711-050		41.5
	Stepan CX 717		34.1
	Stepan CX 717, 5% zinc borate		34.6
	Stepan CX 717, 5% antimony oxide		46.4
	Premi-glas 4100-30,4200-30		20
	Premi-glas 1200-15,1200-22,1200-30, 4000-22,4000-30,4100-15,4100-22		21
	Premi-glas 1100-15,1100-30,4000-15		22
	Premi-glas 1000-30		23
	Premi-glas 1000-15		24
	Premi-glas 2100-30,2200-30,3100-30, 3200-30		28

(Continued)

Table 2.6. Oxygen Index for Various Materials (Continued)

		Thickness in.	Oxygen Index
Polyester,	Premi-glas 2100-22		30 to 32
	Premi-glas 2200-22,3100-15,3200-22		32
	Premi-glas 2100-15		32 to 36
	Premi-glas 3200-15		34
	Premi-glas 2000-30,3000-30		35
	Premi-glas 2200-15		36
	Premi-glas 2200-22		45
	Premi-glas 2000-15,3000-15		55
	Premi-glas 2000-10		60
	maleate, filler, 30% glass		20.9 to 34.1
	isophthalate, filler, 15% glass		27.3 to 36.7
	flame retardant, filler, 15% glass		23.5 to 40.8
	flame retardant, filler, 30% glass		37.0 to 43.4
Alkyd,	Allied Plaskon AMC 417	0.125	40
	Allied Plaskon AMC 422	0.125	51.9
	Allied Plaskon AMC 424	0.125	34.2
	Allied Plaskon AMC 750	0.125	44.1
	Allied Plaskon AMC 794	0.125	30.3
	Allied Plaskon AMC 940	0.125	63.4
	Hooker Durez 24150	0.125	38
	Hooker Durez 24525	0.125	34
	Hooker Durez 24480	0.125	29
	Plenco 1504-B6636		41.0
Silicone rubber,	GE SE 9029		30.3
	GE SE 9036		32.6
	GE SE 9035		25.8
	GE SE 9014		27.9
	GE SE 9044A		30.4
	GE SE 9090		33.7
	GE SE 5537		34.0
	GE 2-2123A		39.2
Rubber,	EPT		21.9
	Hypalon		25.1
	Vulkene		25.1
	Neoprene		26.3
	SBR, coating grade		19.0
	SBR, foam		16.9
	polybutadiene, foam		17.1
	natural, foam		17.2

(Concluded)

Table 2.7. Thermochemical Properties of Various Materials

	Heat of Combustion Kcal/g.mol	Heating Value Btu/lb	Stoichiometric Flame Temperature °C.	°F.
Polyethylene, high density	−312.5	20,050	2120	3850
Polyethylene, low density	−312.0	20,020	2120	3850
Ethylene/propylene polymer, 69/31	−360.8	20,270	2120	3850
Polypropylene, isotactic, syndiotactic	−468.3	20,030	2120	3850
Polypropylene, atactic	−467.8	20,010	2120	3850
Poly-1-butene, isotactic	−625.4	20,060	2120	3850
Polyisobutylene	−628.2	20,150	2130	3870
Poly-1-pentene, isotactic	−780.8	20,040	2120	3850
Poly-3-methyl-1-butene, isotactic	−780.2	20,030	2120	3850
Poly-4-methyl-1-pentene, isotactic	−935.7	20,010	2120	3850
Poly-1,4-butadiene, atactic	−584.3	19,440	2220	4020
Polytetrafluoroethylene	+ 8.01	−144	–	–
Polychlorotrifluoroethylene	− 31.2	482	320	615
Polyvinyl chloride	−268.0	7,720	1960	3550
Polyvinylidene chloride	−232.4	4,315	1840	3340
Polyvinyl fluoride	−238.8	9,180	1710	3100
Polyvinylidene fluoride	−140.3	3,940	1090	2000
Polystyrene, isotactic	−1033	17,850	2210	4010
Polystyrene, atactic, crystal	−1034	17,870	2210	4010
Poly-alpha-methylstyrene	−1196	18,220	2210	4010
Butadiene/styrene (8.58%) copolymer	−604.2	19,300	2220	4020
Butadiene/styrene (25.5%) copolymer	−650.5	19,010	2220	4020
Butadiene/acrylonitrile (37%) copolymer	−512.6	17,180	2190	3970
Polyoxymethylene	−121.4	7,280	2050	3750
Polyoxytrimethylene	−437.6	13,560	2130	3860
Polyethylene oxide	−280.6	11,470	2120	3850
Polypropylene oxide, 27% isotactic	−432.7	13,410	2100	3810
Polypropylene, 100% atactic	−432.3	13,400	2120	3810
Chlorinated polyether	−660.7	7,673	1990	3610

(Continued)

Table 2.7. Thermochemical Properties of Various Materials (Continued)

	Heat of Combustion Kcal/g.mol	Heating Value Btu/lb	Stoichiometric Flame Temperature °C.	°F.
Polycarbonate	−1880	13,310	2190	3980
Polyphenylene oxide	−993.8	14,880	2200	3990
Polypropene sulfone	−462.7	7,850	1970	3570
Poly-1-butene sulfone	−618.9	9,290	2000	3640
Poly-1-hexene sulfone	−913.5	11,310	2040	3710
Polymethyl methacrylate	−637.7	11,470	2070	3760
Phenol-formaldehyde (1:1)	−1496	12,000	1860	3380
Urea-formaldehyde (1:2)	−358.8	7,680	1950	3540
Melamine-formaldehyde (1:3)	−749.3	8,310	1990	3610
Polyurethane, ester-based	−743.1	10,180	2100	3810
Polyester, unsaturated	−723.2	12,810	2250	3910
Epoxy, bisphenol A	−1700	14,430	2220	4030
Polyacenaphthalene	−1429	16,900	2230	4040
Polycarbon suboxide	−224.1	5,940	2260	3910
Polytetrahydrofuran	−592.7	14,790	2120	3850
Polyvinyl alcohol	−263.2	10,760	1980	3600
Poly-beta-propiolactone	−333.3	8,330	2075	3770
Polynitroethylene	−278.7	6,870	2670	4830
Polyacrylonitrile	−408.6	13,860	1860	3380
Cellulose	−1011	7,520	–	–
Paper	–	7,590	–	–
Woodflour	–	8,520	–	–
Wood	–	8,835	–	–
Chipboard (90% woodflour, 10% resin)	–	8,715	–	–
Bituminous coal, med. volatile, W.Va.	–	15,178	–	–
Lignite, Texas	–	11,084	–	–
Peat, Minn.	–	9,057	–	–
Oil shale	–	6,300	–	–
No. 1. fuel oil	–	19,800	–	–
No. 6. fuel oil	–	18,300	–	–

(Concluded)

tion. If the net heat of combustion is positive, then the unit mass is making available an excess of heat to increase the exposure of an adjacent unit mass.

Stage V. **Propagation.** The net heat of combustion of a unit mass, decreased by heat lost to the surroundings and increased by heat supplied from external sources such as an adjacent fire, must be sufficient to bring an adjacent unit mass to the combustion stage, for propagation to result. Where the initial unit mass is at an exposed surface, an adjacent unit mass at the surface is more readily brought to the combustion stage than an adjacent unit mass away from the surface, because the material at the surface is exposed to the external fire source, while the material at the interior is shielded from external heat by the solid residue of the original unit mass, and in addition dissipates heat to material deeper into the interior. Combustion air is also far more easily available to material at the surface than in the interior of the material. For this reason, propagation is often treated as a surface phenomenon. For plastic materials which are applied as an exposed surface over substantial areas, surface flame spread is a realistic measure of propagation.

Where the plastic material is not continuously applied for significant distances, or where it is not applied on or as an exposed surface, surface flame spread becomes less realistic a measure of fire propagation than other flammability characteristics such as heat contribution and combustion products of the same material, and the ease of ignition of other materials in the vicinity.

SECTION 2.3. THE BURNING PROCES ON A MASS SCALE

The burning process in a system, such as a room containing articles associated with typical occupancy, can be considered as occurring in five stages, analogous, but on a larger scale, to the stages described for the unit mass:

State I. **Initial Fire.** The original fire source in a system such as a room can vary widely. It may be a burning match, a cigarette dropped on a fabric, an electrical fire from overheated wiring, or a burning door or window ignited by fire in an adjacent room or structure. In almost all cases, the plastic material in the system is not the source of initial fire. In the initial fire stage, the following properties of the plastic material are relevant:

1. Decomposition temperature and behavior. This characteristic is particularly important if the plastic material is relied upon for structure capabilities (Table 2.2).

2. Ease of ignition. This characteristic is measured by properties such as

flash ignition temperature (if the material is exposed to direct flame), self-ignition temperature (if the material is exposed to heat without flame), and limiting oxygen concentration (if the material is exposed to air mixed with combustion gases from the initial fire).

3. Extent of exposure. Plastic materials presenting an exposed surface are more vulnerable to fire than plastic materials shielded by a coating or covering. Plastics near the ceiling, where the heated gases tend to accumulate, are more vulnerable than plastics near the floor.

4. Extent of involvement. Plastics present in large quantities in the system, such as wall and ceiling material, are more significant than those present in relatively small amounts, such as buttons and utensil handles.

Stage II. **Fire Build-Up.** The heat from the initial fire accumulates in the system, raising the temperature of the materials in the system by conduction, convection, and radiation. A limited amount of fire spread may occur in this stage. In the fire build-up stage, the following properties of the plastic materials involved are relevant:

1. Ease of ignition. Easily ignited materials can contribute to fire build-up.

2. Surface flammability. This is an important characteristic if the material is exposed over significantly large surface areas, such as interior finishes and coverings, but is irrelevant where the material is present at widely separated locations, or in small units.

3. Heat contribution. If the material is ignited, the generation of large amounts of heat during combustion will contribute to fire build-up.

4. Smoke production. The evolution of dense smoke hinders the escape of occupants and the location of the fire for extinguishing efforts.

5. Extent of exposure. This is relevant only if the material is exposed to fire in this stage.

6. Extent of involvement. If the material is present in a small enough quantity, it can not make a significant contribution to fire build-up.

7. Fire gases. Highly toxic combustion gases can present a hazard to the occupants.

Stage III. **Flashover.** The point of flashover is the point at which most of the combustible materials in the system reach their ignition temperatures at essentially the same time, so that the combustible materials seem to burst into flame almost simultaneously. The following properties of the plastic material are relevant in this stage:

1. Ease of ignition. The temperatures present in the system at the flashover stage are probably higher than the ignition temperatures of most plastic materials, so that relative ease of ignition is academic except

in the case of high-temperature polymers.

2. Surface flammability. Although much of the heat transfer leading to flashover occurs through radiation, surface flammability can contribute significantly to flashover.

3. Extent of exposure.

4. Extent of involvement.

Stage IV. **Fully Developed Fire.** Essentially all the combustible materials in the system contribute to the fire in this stage, and fire damage is essentially total for these materials. Once the burning process has reached this stage, containment, rather than extinguishment, becomes critical. The most important factor is the total heat contribution of the materials involved. The following properties of the plastic materials involved are relevant:

1. Extent of involvement. The maximum heat contribution is a function of the unit heat of combustion and the total amount of material.

2. Heat contribution.

3. Smoke production.

4. Fire gases.

Surface flammability is irrelevant in this stage, because fire has reached all combustible surfaces. Extent of exposure is essentially total.

Stage V. **Fire Propagation.** The total fuel contribution of the combustible materials in a system, reduced by the heat required to bring the system to the fully developed fire stage, can provide the initial fire for adjacent systems if the boundaries of the system fail to contain the fire. For this reason, fire endurance is an important characteristic of the system boundaries. A "fire wall" is exactly what the term implies: a wall erected to serve as a barrier to fire propagation.

The following properties of plastic materials are relevant in this stage:

1. Fire endurance.
2. Extent of involvement.
3. Heat contribution.
4. Smoke production.
5. Fire gases.

SECTION 2.4 COMBUSTION AND PEOPLE

Some results of the burning process are obvious threats to human life, wherever occupants are present within the burning system and in adjacent systems, but their relative importance can vary with the conditions of each individual fire. Because there can be such wide variety in types of plastic materials and their extent of exposure and involvement, the hazard of each situation must be judged on an individual basis.

The results of the burning process which can threaten human life can be summarized as the following:

1. Oxygen depletion
2. Flame
3. Heat
4. Toxic gases
5. Smoke
6. Structural strength reduction

1. **Oxygen depletion.** The average human being is accustomed to operating satisfactorily with the usual level of about 21 percent oxygen in the atmosphere. When the oxygen content falls to 17 percent, muscular skill is diminished because of a phenomenon called anoxia. At oxygen levels of 10 to 14 percent, a man is still conscious but exhibits faulty judgement, not obvious to himself. At oxygen levels of 6 to 8 percent, breathing ceases, and death by asphyxiation occurs in 6 to 8 minutes. These responses are summarized in Table 2.8. The excitement and exertion occasioned by a fire tend to increase the oxygen demands of the body, and oxygen deficiency symptoms may appear at higher oxygen levels.

Table 2.8. Response of Humans to Various Concentrations of Oxygen

Concentration Percent	Symptoms
21	Normal concentration in air
17	Respiration volume increased, muscular coordination diminished, more effort required for attention and clear thinking
12 to 15	Shortness of breath, headache, dizziness, quickened pulse, quick fatigue upon exertion, loss of muscular coordination for skilled movements
10 to 14	Faulty judgement
10 to 12	Nausea and vomiting, exertion impossible, paralysis of motion
6 to 8	Collapse and unconsciousness, but rapid treatment can prevent death
6 or below	Death in 6 to 8 minutes
2 to 3	Death in 45 seconds

An oxygen concentration of 10 percent is considered the minimum level for survival. Whether this level is reached, and how rapidly it is reached, varies with each fire, and with location within the system, since it is affected by the concentration of combustibles, rate of burning, volume of the system, and rate of ventilation.

2. **Flame.** Burns can be caused by direct contact with flames or heat radiated from flames. Because it is rarely separated by any appreciable distance from the burning materials, flame, unlike fire gases and smoke, rarely presents a direct threat to occupants of adjacent systems. Since burns can result if skin temperature is held above 150°F. for one second, flame temperatures and their radiant heat may prove immediately or eventually fatal.

3. **Heat.** Unlike direct flame, heat can be a hazard to occupants of adjacent systems in addition to occupants of the burning system. Hot air and gases from a fire, ignoring any effects of oxygen depletion or toxicity, can cause burns, heat exhaustion, dehydration, and blockage of the respiratory tract (edema).

A breathing level temperature of 300°F. is considered to be the maximum value for survival. Breathing level, the distance above the floor, is considered to be about 5 feet, although the 4 foot level is sometimes used when children comprise a significant fraction of the occupants. Temperatures above 150°F. are considered untenable, and temperatures in this range can hold back fire-fighters and keep occupants from entering passages leading to exits.

4. **Fire gases.** The toxicity of some gaseous products of combustion is well known, but the concentration of these gases in an actual fire is not well known, even from simulated conditions in laboratory experiments. What gaseous products of combustion are formed is determined by the chemical composition of the material, the amount of available oxygen, the temperature, and the rate of temperature change.

The hazard of toxic fire gases is often psychological and sometimes overemphasized. Fire fatalities from the inhalation of hot air and gases are a significant proportion of fire deaths. Experimental fires in typical structures have shown that in many cases the minimum survival concentration of oxygen or the maximum survival breathing level temperature was reached before any toxic gases attained a lethal concentration.

Because plastic materials are based on carbon-containing polymers, the gaseous products of combustion include carbon dioxide and carbon monoxide. Carbon dioxide is not generally considered toxic, but it can present a hazard in two ways: First, it represents oxygen depletion. In the complete absence of ventilation, a certain volume of carbon dioxide

represents removal of an equal volume of oxygen. Second, it overstimulates the respiratory system, causing an abnormally high intake of other gases, resulting in toxic or lethal concentrations which would not otherwise have been attained. A carbon dioxide concentration of 1.8 percent is said to increase speed and depth of breathing 50 percent, and a 2.5 percent concentration increases breathing rate 100 percent. The response of humans to carbon dioxide is summarized in Table 2.9.

Table 2.9. Response to Humans to Various Concentrations of Carbon Dioxide

Concentration, ppm	Symptoms
250 to 350	Normal concentration in air
900 to 5000	No effect
5000	TLV and MAK value
18,000	Ventilation increased by 50 percent
25,000	Ventilation increased by 100 percent
30,000	Weakly narcotic, decreasing acuity of hearing, increase in pulse and blood pressure
40,000	Ventilation increased by 300 percent, headache, weakness
50,000	Symptoms of poisoning after 30 minutes, headache, dizziness, sweating
80,000	Dizziness, stupor, unconsciousness
90,000	Distinct dyspnoea, loss of blood pressure, congestion, death within 4 hours
100,000	Headaches and dizziness
120,000	Immediate unconsciousness, death in minutes
200,000	Narcosis, immediate unconsciousness, death by suffocation

Carbon monoxide ranks first as a cause of fire deaths because it is always formed by an uncontrolled accidental fire, in which oxygen concentrations and availability are never ideal for complete oxidation to carbon dioxide. The maximum concentration of carbon monoxide for survival is considered to be 1.28 percent. At this or higher levels, a person will become unconscious after two or three breaths and probably die in one to three minutes. The response of humans to various levels of carbon monoxide is summarized in Table 2.10. Variables such as physical condition and amount of exertion affect the amount of carbon monoxide which can be tolerated.

Carbon monoxide combined with the hemoglobin in the blood to form carboxyhemoglobin, displacing oxygen in the blood and leading to

Table 2.10. Response of Humans to Various Concentrations of Carbon Monoxide

Concentration, ppm	Symptoms
25	TLV for conditions of heavy labor, high temperatures, and decreased air pressure
50	TLV and MAK value
100	No poisoning symptoms even for long periods of time, allowable for several hours
200	Headache after 2 to 3 hours, collapse after 4 to 5 hours
300	Headache after 1.5 hours, distinct poisoning after 2 to 3 hours, collapse after 3 hours
400	Distinct poisoning, frontal headache, and nausea after 1 to 2 hours, collapse after 2 hours, death after 3 to 4 hours
500	Hallucinations felt after 30 to 120 minutes
800	Collapse after 1 hour, death after 2 hours
1000	Difficulty in ambulation, death after 2 hours
1500	Death after 1 hour
2000	Death after 45 minutes
3000	Death after 30 minutes
8000 or above	Immediate death by suffocation
12,800	Unconsciousness after 2 to 3 breaths, death in 1 to 3 minutes

anoxia and death. The response of humans to carboxyhemoglobin is summarized in Table 2.11.

The response of animals to various levels of carbon oxides is summarized in Table 2.12.

Some gases such as sulfur dioxide and ammonia are extremely irritating at concentrations well below lethal levels, and a person will normally leave the system or don protective breathing apparatus, if he is able, before suffering serious effects. Hydrogen sulfide is identified by its "rotten egg" odor, but continued exposure or concentrations above 0.2 percent paralyze the sense of smell. Hydrogen cyanide has a characteristic bitter almond odor which may be masked by other odors. Nitrogen dioxide is identified by its reddish-brown color, but it tends to anesthetize the throat and its toxic effect is often delayed. Fatalities may occur hours or days after exposure although the exposed persons may show no immediate ill effects.

The response of humans to various toxic gases is summarized in Table 2.13. The response of animals to some of these gases is summarized in Table 2.14.

Table 2.11. Response of Humans to Various Concentrations of Carboxyhemoglobin

Concentration, percent	Symptoms
0 to 10	No signs or symptoms
10 to 20	Tightness across forehead, possible slight headache, dilation of cutaneous blood vessels
20 to 30	Headache, throbbing in temples
30 to 40	Severe headache, weakness, dizziness, dimness of vision, nausea, vomiting, collapse
40 to 50	Same as above, increase in pulse and breathing rate, greater possibility of collapse, asphyxiation
50 to 60	Same as above, coma, intermittent convulsions, and Cheyne-Stokes respiration
60 to 70	Coma, intermittent convulsions, depressed heart action and respiratory rate, possible death
70 to 80	Weak pulse, slowing of respiration leading to death within hours
80 to 90	Death in less than 1 hour
90 to 100	Death in a few minutes

Table 2.12. Response of Animals to Various Concentrations of Carbon Oxides

Carbon Oxide and Concentration, ppm	Effects
Carbon dioxide	
400,000	Lethal to mice after 4 hours
Carbon monoxide	
1250	Lethal to mice after 4 hours
1500	25 percent carboxyhemoglobin in mice after 5 minutes
2100	25 percent carboxyhemoglobin in rates after 5 minutes
4670	LC50 for rats in 60 minutes
5000	Minimum lethal dose for rats in 30 minutes
5500	LC50 for rats in 30 minutes
6100	LC50 for rats in 20 minutes
8800	LC50 for rats in 10 minutes

Table 2.13. Response of Humans to Various Concentrations of Materials

Compound and Concentration, ppm	**Symptoms**
Hydrogen cyanide (HCN)	
0.2 to 5.1	Threshold of odor
10	TLV and MAK value
18 to 36	Slight symptoms, headache, after several hours
45 to 54	Tolerated for ½ to 1 hour without difficulty
100	Fatal after 1 hour
110 to 135	Fatal after ½ to 1 hour, dangerous to life
135	Fatal after 30 minutes
181	Fatal after 10 minutes
280	Immediately fatal
Acetonitrile	
40	TLV and MAK value, odor detectible, no symptoms
80	No symptoms after 4 hours
160	Slight feeling of bronchial tightness after 4 hours
Acrylonitrile	
20	TLV and MAK value
Acetone	
0.5 to 1000	Threshold of odor, quick adaption
1000	TLV and MAK value
2000	No symptoms on workers exposed over many years
5000	Induces first narcotic symptoms
9300	Irritation of throat after 5 minutes
Propane	
1000	MAK value
Butane	
1000	MAK value
Hexane	
500	TLV and MAK value
Octane	
500	TLV and MAK value
Octane	
400	TLV value
500	MAK value
1,3-Butadiene	
1000	TLV and MAK value

(Continued)

Table 2.13. Response of Humans to Various Concentrations of Materials (Continued)

Compound and Concentration, ppm	Symptoms
Hydrogen chloride (HCl)	
1 to 5	Limit of detection by odor
5	TLV and MAK value
5 to 10	Mild irritation of mucous membranes
35	Irritation of throat on short exposure
50 to 100	Barely tolerable
1000	Danger of lung edema after merely short exposure
Hydrogen fluoride (HF)	
3	TLV and MAK value
3 to 5	Redness of skin, irritation of nose and eyes after one week exposure
32	Irritation of eyes and nose
60	Itching of skin, irritation of respiratory tract from ex posure of 1 minute
120	Conjunctival and respiratory irritation just tolerable for 1 minute
50 to 100	Dangerous to life after a few minutes
Ammonia (NH_3)	
1 to 50	Detectable odor
25	TLV value
50	MAK value
57 to 72	Respiration not significantly changed
96	Slight irritation of nose, throat, and eyes
100	Working possible, adaptation
200	Irritation of the mucous membranes
500 to 1000	Strong irritation of upper respiratory tract
2000	Fatal
Nitrogen dioxide (NO_2)	
5	TLV and MAK value, threshold of perception by odor
10 to 20	Mildly irritant to eyes, nose, and upper respiratory tract
25 to 38	No adverse effects in workers exposed over several years
50	Distinct irritation
80	Tightness of chest after 3 to 5 minutes
90	Pulmonary edema after 30 minutes
100 to 200	Very dangerous within 30 to 60 minutes
250	Death after a few minutes

(Continued)

Table 2.13. Response of Humans to Various Concentrations of Materials (Continued)

Compound and Concentration, ppm	Symptoms
Sulfur dioxide (SO_2)	
3 to 5	Odor threshold
5	TLV and MAK value
8 to 12	Slight irritation of eyes and throat, resistance of air tracts
20	Coughing and eye irritation
30	Immediate strong irritation, remains very unpleasant
100 to 250	Dangerous to life
600 to 800	Death in a few minutes
Hydrogen sulfide (H_2S)	
10	TLV and MAK value
20 to 30	Conjunctivitis
50	Objection to light after 4 hours, lachrymation
50 to 500	Irritation of respiratory tract
150 to 200	Objection to light, irritation of mucous membranes, headache
200 to 400	Slight symptoms of poisoning after several hours
250 to 600	Pulmonary edema and bronchial pneumonia after prolonged exposure
500 to 1000	Systemic poisoning, painful eye irritation, vomiting
1000	Immediate acute poisoning
1000 to 2000	Lethal after 30 to 60 minutes
over 2000	Acute lethal poisoning
Acetic acid	
10	TLV value, not irritating
20 to 30	No danger to workers exposed for 7 to 12 years
25	MAK value
60	Slight irritation of respiratory tract, stomach, and skin
800 to 1200	Cannot be tolerated for more than 3 minutes
Ethyl acetate	
0.2	Limit of perception of odor
200	Strong odor perceived
350	Irritation of nose and eyes
400	TLV and MAK value
700	Narcotic effects without fainting
3800	Distinct narcosis with fainting and significant irritation

(Continued)

Table 2.13. Response of Humans to Various Concentrations of Materials (Continued)

Compound and Concentration, ppm	Symptoms
Butyl acetate	
150	TLV value
200	MAK value, irritation of upper respiratory tract, nose, eyes, and throat
300	Strong irritation effects
900	Strong irritation, appearance of narcotic effects with sensation of vertigo
1800	Deep narcosis with vertigo and unconsciousness
Toluene diisocyanate (TDI)	
0.01	No irritation, no odor for 30 minutes
0.018 to 0.02	Odor threshold
0.02	TLV and MAK value
0.05	Slight irritation of the eyes
0.1	Tolerable irritation of eyes, nose and throat
0.5	Heavy irritation of eyes, nose and throat
1.3	Heavy irritation, coughing, spasms of the bronchi, tracheitis lasting several hours after exposure
Formaldehyde	
0.05 to 1.0	Threshold of odor
0.08 to 1.6	Slight irritation of eyes and nose
0.25 to 1.6	Threshold of irritation of eyes
0.5	Threshold of irritation of throat
1.0	MAK value
2.0	TLV value
10	Conjunctivitis, rhinitis, and pharyngitis within a few minutes
10 to 15	Dyspnoea, cough, pneumonia, bronchitis
over 50	Necrosis of mucous membranes, spasm of larynx, edema of lungs
Acetaldehyde	
0.07 to 0.21	Threshold of odor
25 to 50	Transient slight irritation of eyes after 15 minutes
100	TLV value
134	Slight irritation of respiratory tract after 30 minutes
200	MAK value, irritation of nose and throat
Acrolein	
0.1	TLV and MAK value
0.805	Lachrymation, irritation of mucous membranes

(Continued)

Table 2.13. Response of Humans to Various Concentrations of Materials (Continued)

Compound and Concentration, ppm	Symptoms
Acrolein	
1.0	Immediately detectable, irritation
5.5	Intense irritation
10 and over	Lethal in a short time
24	Unbearable
Benzene	
25	TLV value
500	Slight irritation
1500 to 4000	Dangerous to life after several hours
8000	Fatal after 30 to 60 minutes
20,000	Fatal after 5 minutes
Toluene	
100	TLV value
200	MAK value
190 to 380	No complaints
500 to 1000	Headache, nausea, momentary loss of memory, anorexia, irritation of eyes
1000 to 1500	Palpitation, extreme weakness, loss of coordination, impairment of reaction time
2000 to 2500	Dizziness, nausea, narcosis after 3 hours
10,000	Immediately fatal
Styrene	
60	Threshold of odor, no irritation
100	TLV and MAK value, strong odor, tolerable
200 to 400	Intolerable odor
216	Unpleasant subjective symptoms
376	Definite signs of neurological impairment
600	Irritation of eyes
800	Immediate irritation of eyes and throat, somnolence, weakness
over 800	Nausea, vomiting, and total weakness

5. **Smoke.** The principal hazard of smoke is that it may hinder the escape of occupants and the entry of fire-fighters seeking to locate and extinguish the fire. It can contribute to panic conditions because of its blinding and irritating effects. In many cases, smoke reached untenable levels in exitways before temperature reached untenable levels.

The amount of light absorption by photometer methods is not necessarily indicative of the degree of sight obscuration. The size and

Table 2.14. Response of Animals to Various Concentration of Materials

Compound and Concentration, ppm	Effects
Hydrogen cyanide (HCN)	
50	Minimum lethal dose for rats in 30 minutes
56	Symptoms in cats after 60 minutes
50 to 90	Symptoms in rats after 30 minutes
100	May be fatal to rats after 30 minutes
110	Fatal to dogs and cats after 30 minutes
120	LC50 for rats in 60 minutes
125	Severe symptoms in monkeys after 12 minutes
142	LC50 for rats in 30 minutes
200	LC50 for rats in 30 minutes
315	Fatal to dogs and cats after 5 minutes
323	LC50 for mice in 5 minutes
503	LC50 for rats in 5 minutes
Acetone	
20,000	Narcosis in guinea pigs after 8 to 9 hours
20,256	Narcosis in mice after 1 hour
46,000	Narcosis in rats after 2 hours, lethal to mice in 1 hour
50,000	Lethal to guinea pigs in 3 hours
126,000	Lethal to rats in 2 hours
Hydrogen chloride (HCl)	
50	Tolerable to monkeys for 6 hours daily over 20 days
300	Mild corneal damage in guinea pigs after 6 hours
3200	No mortality in mice after 5 minutes
4300	Lung edema in rabbits, death after 30 minutes
13,745	LC50 for mice in 5 minutes
30,000	No mortality in rats after 5 minutes
40,989	LC50 for rats in 5 minutes
Hydrogen fluoride (HF)	
7	No effect on various species after daily exposure
40	Guinea pigs and rabbits survived for 41 hours
300	Death in guinea pigs and rabbits after 2 hours
2430	Highest concentration without mortality in mice after 5 minutes
6247	LC50 for mice in 5 minutes
10,000	Highest concentration without mortality in rats after 5 minutes
18,200	LC50 for rats in 5 minutes

(Continued)

Table 2.14. Response of Animals to Various Concentrations of Materials (Continued)

Compound and Concentration, ppm	Effects
Carbonyl fluoride	
360	LC50 for rats in 1 hour
Tetrafluoroethylene	
40,000	LC50 for rats in 4 hours
Ammonia	
480 to 570	Irritation in various species after 4 hours
800 to 1070	Strong irritation in various species after 7 hours
820 to 1430	Dyspnoea, tracheitis in various species after 5 to 7 hours
2000	No deaths in rats after 4 hours
3420	Minimum lethal concentration for mice after 2 hours
4000	Fatal to rats after 4 hours
4760	LC50 for mice in 2 hours
10,930	LC50 for rats in 2 hours
20,000 to 30,000	Immediately fatal to various species after a few minutes
Nitrogen dioxide	
50	No mortality in mice after 4 hours
100	Lethal to mice after 4 hours
260	Highest concentration without mortality in rats and mice after 5 minutes
831 to 832	LC50 for rats in 5 minutes
1880	LC50 for mice in 5 minutes
Sulfur dioxide	
400	Most species survive after 1 to 5 hours
600 to 800	Lethal to some species
Toluene diisocyanate	
9.7	LC50 for mice in 4 hours
11.0	LC50 for rabbits in 4 hours
12.7	LC50 for guinea pigs in 4 hours
13.9	LC50 for rats in 4 hours
Formaldehyde	
250	LC50 for rats in 30 minutes
830	LC50 for rats in 4 hours
Acetaldehyde	
3000 to 20,000	Edema of lungs and narcotic effects in animals

(Concluded)

distribution of smoke particles affect the optical absorption level at which sight obscuration is essentially total, as does the degree of irritation to the observer. Smoke density is influenced by the rate of burning and degree of ventilation, and is generally inversely proportional to the degree of ventilation.

6. **Structural strength reduction.** The failure of structural components through heat damage or burning can present a serious hazard. Perhaps the most dramatic examples are the collapse of weakened floors under the weight of firemen, and the collapse of walls and roofs on persons beneath them.

CHAPTER 3

Fire Response Characteristics

Fire response characteristics are those properties which describe the response of a material when exposed to fire. For both convenience and convention, fire response characteristics are generally considered to include the following:

1. Smolder susceptibility
2. Ignitability
3. Flash-fire propensity
4. Flame spread
5. Heat release
6. Fire endurance
7. Ease of extinguishment
8. Smoke evolution
9. Toxic gas evolution

SECTION 3.1. TYPES OF FIRE RESPONSE CHARACTERISTICS

Fire response characteristics can be classified according to the type of measurements used to describe them. Three types of measurements are used:

1. Thermophysical: smolder susceptibility, ignitability, flash-fire propensity, flame spread, heat release, fire endurance, and ease of extinguishment.
2. Optical: smoke evolution.

3. Biological: toxic gas evolution

In some situations, thermogravimetric measurements are substituted for optical measurements in describing smoke evolution, and chemical or biochemical measurements are substituted for biological measurements in describing toxic gas evolution.

SECTION 3.2. SMOLDER SUSCEPTIBILITY

Smolder has been defined as to burn without flame, and smoldering has been defined as combusting without flame, usually with incandescence and moderate smoke. Smoldering can continue for relatively long periods of time, during which it may or may not kindle into flame. Smoldering combustion can be considered to be the slow propagation of a combustion wave through porous fuel, characterized by relatively low temperatures and incomplete oxidation controlled by the diffusion rate of oxygen through the porous fuel.

Smoldering combustion generally requires a porous fuel, and charring appears to be a prerequisite for smoldering. Cellulosic materials such as wood, cotton, and rayon often tend to form char and become prone to smoldering; smoldering has occurred in cellulose fiber insulation and cellulose fiberboard. Some synthetic materials also tend to form char, either by design or by treatment, and these can become prone to smoldering; smoldering has occurred in certain phenolic and polyurethane foams.

Smolder susceptibility test data on some materials are presented in Table 3.1.

SECTION 3.3. IGNITABILITY

Ignitability has been defined as the ease of ignition, especially by a small flame or spark. Ignition has been defined as the initiation of combustion.

The thermal exposure causing ignition is a combination of heat flux and time. The higher the heat flux, the shorter the time before ignition occurs for a given material. Thermal insulation materials permit less heat to flow from the exposed surface to the interior of the material, and as a result reach surface ignition more rapidly than less effective insulators.

Ignitability test data on some materials are presented in Table 3.2.

Table 3.1. Smolder Susceptibility of Materials as Measured by the California Bureau of Home Furnishings

	Number of Fabrics Tested	Number of Fabrics Igniting	Percent Igniting
Fabric on fiberglass board			
100% cellulosic fabrics	42	42	100
Cellulosic/thermoplastic fabric blends	28	15	53.6
Miscellaneous fabrics	7	1	14.3
100% thermoplastic fabrics	11	0	0
Fabric on polyurethane foam (no FR)			
100% cellulosic fabrics	42	17	40.5
Cellulosic/thermoplastic fabric blends	28	7	25.0
Miscellaneous fabrics	7	1	14.3
100% thermoplastic fabrics	11	0	0
Fabric on polyurethane foam (FR, 1)			
100% cellulosic fabrics	42	36	85.7
Cellulosic/thermoplastic fabric blends	28	15	53.6
100% thermoplastic fabrics	11	0	0
Miscellaneous fabrics	7	0	0
Fabric on polyurethane foam (FR, 2)			
100% cellulosic fabrics	42	16	38.1
Cellulosic/thermoplastic fabric blends	28	7	25.0
100% thermoplastic fabrics	11	0	0
Miscellaneous fabrics	7	0	0
Fabric on polyurethane foam (HR)			
100% cellulosic fabrics	42	35	83.3
Miscellaneous fabrics	7	4	57.1
Cellulosic/thermoplastic fabric blends	28	14	50.0
100% thermoplastic fabrics	11	0	0
Fabric on cotton batting (no FR)			
100% cellulosic fabrics	42	42	100
Cellulosic/thermoplastic fabric blends	28	23	82.1
Miscellaneous fabrics	7	5	71.4
100% thermoplastic fabrics	11	1	9.1
Fabric on cotton batting (FR)			
100% cellulosic fabrics	42	32	76.2
Cellulosic/thermoplastic fabric blends	28	12	42.9
Miscellaneous blends	7	1	14.3
100% thermoplastic fabrics	11	0	0

(Continued)

Table 3.1. Smolder Susceptibility of Materials as Measured by the California Bureau of Home Furnishings (Continued)

	Number of Fabrics Tested	Number of Fabrics Igniting	Percent Igniting
Fabric on non-resinated polyester batting			
100% cellulosic fabrics	42	14	33.3
Cellulosic/thermoplastic fabric blends	28	2	7.1
100% thermoplastic fabrics	11	0	0
Miscellaneous fabrics	7	0	0
Fabric on resinated polyester batting			
100% cellulosic fabrics	42	8	19.0
Cellulosic/thermoplastic fabric blends	28	1	3.6
100% thermoplastic fabrics	11	0	0
Miscellaneous fabrics	7	0	0
Fabric on 70/30 cotton/polyester batting			
100% cellulosic fabrics	42	33	78.6
Cellulosic/thermoplastic fabric blends	28	9	32.1
Miscellaneous fabrics	7	1	14.3
100% thermoplastic fabrics	11	0	0
Fabric on neoprene foam			
100% cellulosic fabrics	42	39	92.9
Cellulosic/thermoplastic fabric blends	28	11	39.3
Miscellaneous fabrics	7	1	14.3
100% thermoplastic fabrics	11	0	0
Fabric on neoprene foam interliner and cotton batting (no FR)			
100% cellulosic fabrics	42	8	19.0
Miscellaneous fabrics	7	1	14.3
Cellulosic/thermoplastic fabric blends	28	1	3.6
100% thermoplastic fabrics	11	0	0

(Concluded)

SECTION 3.4. FLASH-FIRE PROPENSITY

Flash fires are a special form of fire hazard which combine the different aspects of ignitability, flammability, heat release, and flame spread. A flash fire has been defined as a fire (flame front) which propagates through a fuel-air mixture as a result of the energy release from the combustion of that fuel, having required only an ignition source. A flash

Table 3.2. Ignitability Characteristics of Materials as Measured by the USF Ignitability Test

Material	Heat Flux, W/cm²	Ignition Time, sec		
		5.8	8.1	10.5
Aspen poplar, ¾ in		42	17	7
Beech, ¾ in		79	20	11
Yellow birch, ¾ in		*	22	11
Red oak, ¾ in, 1		42	21	14
Red oak, ¾ in, 2		55	23	17
Western red cedar, ¾ in		42	8	6
Douglas fir, ¾ in		44	23	11
Western hemlock, ¾ in		44	21	10
Eastern white pine, ¾ in		30	12	7
Southern yellow pine, ¾ in		39	16	9
Hemlock, untreated, ¾ in		43	16	
Hemlock, treated, ¾ in		74	43	
Cellulose fiberboard, core board, ½ in		11	5	3
Fiberboard soundstop, ½ in		15	6	4
Fiberboard sheathing, asphalt impregnated, ½ in		17	7	5
Medium density hardboard, ½ in		34	21	14
Hardboard, 3/8 in		84	26	17
Chipboard, 1/8 in		77	23	14
Hardboard, unfinished, ¼ in		222	22	12
Lauan hardwood plywood, unfinished, AD, ¼ in		229	17	10
Polymethyl methacrylate, clear, 1/8 in		115	33	31
Polystyrene, white, 1/8		108	35	31
Polyisocyanurate rigid foam, ½ in		*	*	56
Polyvinyl chloride flooring, 1/8 in		269	95	78
Linoleum, 1/8 in		59	54	12
Wool carpet, latex foam rubber backing		26	9	6
Polyester carpet, resin backing		79	42	28
Nylon carpet, latex foam rubber backing		29	17	12
Acrylic carpet, vegetable fiber backing		38	15	9
Wool fabric, FR		104	62	15
Wool/nylon fabric, FR		20	16	8
Rayon fabric, FR		12	9	5
Cotton fabric		10	6	3
Nylon fabric		22	26	7

*No ignition

fire has also been defined as a fire that spreads with extreme rapidity, such as one that races over flammable liquids and through gases. Extreme rapidity of flame propagation seems to be generally considered as characteristic of a flash fire.

It should be emphasized that a flash fire is not the same as flashover. Flashover has been defined as a stage in the development of a contained fire in which all exposed surfaces reach ignition temperature more or less simultaneously and fire spreads throughout the space and flames appear on all surfaces.

Flash fires tend to occur when combustible gases are evolved as a result of thermal exposure, escape burning at the original point of exposure, and accumulate to form fuel-air mixtures which then contact an ignition source.

Flash-fire test data on some materials are presented in Table 3.3.

Table 3.3. Flash-Fire Propensity of Materials as Measured by the USF Flash-Fire Screening Test

Material	**Height of Flash Fire inches**	**Time to Flash Fire seconds**
Hardwoods		
Aspen poplar	20 ± 6	31 ± 5
Beech	15 ± 3	36 ± 12
Yellow birch	22 ± 4	28 ± 4
Red Oak	26 ± 0	34 ± 2
Softwoods		
Western red cedar	26 ± 0	29 ± 8
Douglas fir	26 ± 0	39 ± 11
Western hemlock	22 ± 3	25 ± 2
Eastern white pine	23 ± 6	26 ± 5
Southern yellow pine	26 ± 0	33 ± 8
Cellulosic board		
Fiberboard, core board	9 ± 3	35 ± 14
Fiberboard soundstop	10 ± 1	27 ± 5
Fiberboard sheathing, asphalt-imp	12 ± 4	41 ± 10
Medium density hardboard	8 ± 1	29 ± 1
Hardboard	13 ± 6	29 ± 13
Chipboard	13 ± 3	35 ± 10

(Continued)

Table 3.3. Flash-Fire Propensity of Materials as Measured by the USF Flash-Fire Screening Test (Continued)

Material	Heigh of Flash Fire inches	Time to Flash Fire seconds
Solid plastics		
Polyethylene, 1	26 ± 0	115 ± 20
Polyethylene, 2	26 ± 0	118 ± 7
Polystyrene	8 ± 4	160 ± 37
Polymethyl methacrylate	26 ± 0	61 ± 18
ABS, 1	7 ± 2	127 ± 32
ABS, 2	10 ± 2	189 ± 7
Polycarbonate, 1	6 ± 3	97 ± 13
Polycarbonate, 2	none	none
Nylon 6	26 ± 0	55 ± 7
Nylon 6/6	26 ± 0	97 ± 37
Nylon 6/10	22 ± 8	77 ± 10
Polyvinyl chloride (3 samples)	none	none
Chlorinated polyvinyl chloride	none	none
Polyphenylene oxide, modified	none	none
Polyphenylene sulfide	none	none
Polyether sulfone	none	none
Polyaryl sulfone	none	none
Rigid foam plastics		
Polyester	22 ± 7	91 ± 8
Polyethylene	26 ± 0	141 ± 4
Polyurethane, no FR	3 ± 1	92 ± 4
Polyurethane, 10% FR (Cl)	3 ± 1	86 ± 11
Polyurethane, 7% FR (P)	4 ± 1	98 ± 6
Polyurethane, high density, FR, 1	4 ± 1	79 ± 7
Polyurethane, high density, FR, 2	6 ± 2	48 ± 8
Polyisocyanurate, urethane-mod., 1	2 ± 1	72 ± 5
Polyisocyanurate, urethane-mod., 2	none	none
Polymethacrylimide	8 ± 1	69 ± 4
Polybismaleimide	6 ± 2	50 ± 2
Polyimide, modified	6 ± 3	44 ± 1
Polystyrene	3 ± 2	203 ± 10
Polyvinyl chloride	none	none

(Continued)

Table 3.3. Flash-Fire Propensity of Materials as Measured by the USF Flash-Fire Screening Test (Continued)

Material	Height of Flash Fire inches	Time to Flash Fire seconds
Flexible foam plastics		
Polyurethane (18 samples)	26 ± 1	44 ± 4
Polyurethane, F6	14 ± 7	33 ± 12
Polyurethane, H1	none	none
Polychloroprene (2 samples)	none	none
Upholstery fabrics		
Cotton (4 samples)	20 ± 3	36 ± 16
Rayon (3 samples)	20 ± 7	37 ± 12
Nylon (3 samples)	21 ± 6	83 ± 8
Polypropylene (3 samples)	25 ± 2	106 ± 13
Cushioning materials		
Cotton batting, no FR	14 ± 3	25 ± 3
Cotton batting, 10% boric acid	8 ± 4	29 ± 4
Excelsior	11 ± 3	21 ± 4
Sisal	23 ± 6	23 ± 4
Kapok	13 ± 2	27 ± 5
Loose fill insulation		
Cellulose, untreated	22 ± 3	18 ± 3
Cellulose, 36.3% 5-mol borax	6 ± 2	20 ± 2
Cellulose, 36.7% boric acid	1 ± 0	18 ± 1

(Concluded)

SECTION 3.5. FLAME SPREAD

Flame spread has been defined as the progress of flame over a surface. Because flame spread is a surface phenomenon, it is critically affected by whether combustible gases are evolved at the surface, or are evolved in the interior of the material and escape at the surface.

Flame spread requires that successive sections of surface be brought to the ignition temperature as a result of heat flux from the advancing flame. The ignitability characteristics of the material therefore have a direct bearing on flame spread. Thermal insulation materials tend to reach surface ignition more rapidly, and as a result tend to exhibit more rapid flame spread.

Flame-spread test data on some materials are presented in Table 3.4.

Table 3.4. Flame Spread and Smoke Characteristics of Materials as Measured by the ASTM E 84 Test

Material	Flame Spread Classification	Smoke Developed Classification
Red oak, untreated	100	100
Red oak, treated	25 to 50	0 to 45
Oak, treated	35	25
Aspen, treated	30	20
Yellow poplar, treated	25	30
Ash, treated	60	5
Birch, treated	15 to 30	0 to 15
Lauan, treated	15 to 20	5 to 25
Soft maple, treated	20 to 30	0 to 15
Basswood, treated	25 to 35	25 to 45
Cottonwood, treated	25	20 to 60
Virola, treated	15	10
Northern pine, treated	20	20 to 40
Ponderosa pine, treated	25 to 35	30 to 70
Jack pine, treated	25 to 55	15 to 35
Yellow pine, treated	25	25 to 30
Southern yellow pine, treated	10 to 15	0 to 10
Douglas fir, treated	15 to 25	0 to 15
White fir, treated	20	0
Hemlock, treated	15 to 25	0 to 10
Western hemlock, treated	20	0
Redwood, treated	20	0
Western red cedar, treated	45	25 to 40
Molded plastic	10 to 200	65 to over 500
Reinforced plastic	15 to 160	140 to over 500

SECTION 3.6. HEAT RELEASE

The amount of heat released from a burning material, and the rate at which that heat is released, influence the temperature of the fire environment and the rate of fire spread. Heat release rate has been defined as the quantity of heat liberated during complete combustion, usually expressed per unit time and per unit quantity of material.

Heat release test data on some materials are presented in Table 3.5.

SECTION 3.7. FIRE ENDURANCE

Fire endurance has been defined as a measure of the elapsed time during which a material or product maintains its design integrity under specified conditions of test and performance. This measure of fire

Table 3.5. Heat Release Characteristics of Materials as Measured by the OSU Release Rate Test

Material	Orientation V-vertical H-horizontal	Applied Heat Flux W/cm²	Maximum Heat Release Rate Btu/ft²min	Total Heat Release Btu/ft²	
				3 min	10 min
Oak, 1 in	V	1.0	300	250	750
		2.0	420	800	3,500
		2.5	550	1,000	4,000
Pine, 1 in	V	1.5	600	900	6,000
		2.5	800	1,200	8,000
Red oak, flooring	H	1.0	300	500	800
		2.0	450	900	3,500
		2.75	600	1,200	4,500
Hardboard, 0.25 in	V	1.0	800	400	4,500
		2.0	1,200	800	5,500
Particle board, 0.5 in	V	1.0	400	70	1,800
		2.5	800	1,800	5,000
Particle board, FR, 0.5 in	V	1.5	20	0	50
		2.5	350	200	1,200
Fiberboard, low density, 0.625 in	V	1.0	550	1,000	1,500
		2.0	700	1,500	1,600
Exterior plywood, 0.5 in	V	1.0	400	700	1,200
		2.0	600	1,000	3,000
Polypropylene sheet, 0.125 in	H	1.0	2,200	250	7,000
Polyvinyl chloride, rigid, pipe	V	1.0	20	0	50
		2.6	300	100	1,500
Polyvinyl chloride, flexible, sheet, 90 mil	V	2.0	300	500	2,000
		2.0	500	1,000	2,000
Polystyrene, light diffuser	H	0	500	75	2,400
		1.0	750	250	2,500
ABS, sheet, 0.125 in	H	0	600	75	800
		1.0	1,800	600	1,800
		2.5	1,800	1,650	2,400
Polymethyl methacrylate, light diffuser	H	0	500	175	3,000
		1.0	800	300	3,500

Table 3.5. Heat Release Characteristics of Materials as Measured by the OSU Release Rate Test (Continued)

Material	Orientation V-vertical H-horizontal	Applied Heat Flux W/cm²	Maximum Heat Release Rate Btu/ft²-min	Total Heat Release Btu/ft² 3 min	10 min
Polysulfone, sheet, 95 mil	H	1.0	0	0	0
		2.5	450	50	1,500
Polyurethane rigid foam, 1 in	V	0	300	500	500
		1.0	500	500	500
Polyurethane rigid foam, FR, 1 in	V	0.5	0	0	0
		1.0	300	350	350
		2.5	400	650	650
Polyisocyanurate rigid foam, 1 in	V	1.0	0	0	0
		2.75	400	100	150

(Concluded)

resistance is usually expressed in terms of hours or fractions of hours of time during which a material exposed to a specified fire environment must continue exhibit a specified level of performance.

Fire endurance data are generally obtained only for complete systems such as floor and wall constructions.

SECTION 3.8. EASE OF EXTINGUISHMENT

The ease with which fire can be extinguished varies with the material which is serving as the fuel for the fire, and some burning materials are more difficult to extinguish than others. One measure of ease of extinguishment is the oxygen concentration required to support continued burning, and the oxygen index is useful for this purpose.

Oxygen index data on some materials are presented in Table 2.6.

SECTION 3.9. SMOKE EVOLUTION

Smoke has been defined as a visible, nonluminous, airborne suspen-

Table 3.6. Smoke Density from Various Materials (National Bureau of Standards Test)

	Thickness	Maximum Specific Optical Density, Dm		Time to Ds = 16 minutes	
	mils	nonflaming	flaming	nonflaming	flaming
Polyethylene,					
UCC DXM-100	250	526		4.52	
UCC DFDA-6311	125	719	387	2.74	1.09
UCC DHDA-1811	125	739	375	3.76	1.09
UCC DMDA-7075	125	357	280	3.32	1.09
unidentified		468	150	5.5	3
	125		68		3.2
Polypropylene,					
UCC JMD-8500	250	780	119	3.00	4.18
40% glass	100	691	428	2.13	1.57
rug	180	456	110	2.3	1.7
rug, burlap back	220	621	292		
Polytetrafluoroethylene		0	53	NR	11
Tetrafluoroethylene-vinylidene fluoride copolymer	71	75	109	2.5	1.2
Polyvinyl fluoride	2	1	4	NR	NR
Polyvinyl chloride,					
UCC QCA-2460	15	11	98	NR	0.45
	20	23	153	2.66	0.28
	40	139	326	1.16	0.40
UCC QYTQ	250	315	780	3.25	0.49
rigid, filled	250	490	530	1.6	0.5
rigid, unfilled	125	270	525	2.1	0.5
	250	470	535	2.1	0.6
unidentified	10		100		
	15		210		
	30	120	430		
unidentified	250	300	660	3.9	0.8
fabric	26	261	198	1.4	0.3
Polyvinyl butyral,					
UCC XYHL	250	33	31	4.71	6.50
Polystyrene,					
UCC SMD-3500	250	395	780	4.00	0.63
unidentified		460	468	4.0	1.2
unidentified	125		660		0.6
unidentified	250	372	660	7.3	1.3
Styrene-acrylonitrile,					
UCC C-11	250	389	249	4.13	1.11
40% glass	100	687	684	2.28	0.47

(Continued)

Table 3.6. Smoke Density from Various Materials (National Bureau of Standards Test) (Continued)

	Thickness	Maximum Specific Optical Density, Dm		Time to Ds =16 minutes	
	mils	nonflaming	flaming	nonflaming	flaming
Acrylonitrile-butadiene-styrene (ABS),					
Cycolac	250	780	780	2.98	0.57
25/10/65, sheet	45	76	660	4.3	0.6
	80	167	660	3.5	0.7
unidentified	22	188			
	32	220	450		
unidentified	46	71	660	4.8	0.6
unidentified, sheet	70	152	660	1.7	0.4
Polymethylmethacrylate,					
FR UV-Abs	250	380	480	3.8	1.8
HR UV-Abs	250	195	90	6.3	2.2
HR UV-Trans	250	190	140	6.5	2.3
unidentified, sheet	180	203	383	6.0	1.9
unidentified	219	156	107	9.2	2.6
unidentified, sheet	230	304	660	4.5	1.4
unidentified, sheet	500	328		9.0	
unidentified	250	142	81		
Acrylic, rug, Acrilan	300	319	159	1.5	0.6
carpet, 82 oz/sq.yd	375	470	220		
Modacrylic,					
fabric	24	54	21	0.5	4.0
fabric, 5.8 oz/sq.yd	13	41	39	0.8	0.4
fabric, 5.9 oz/sq.yd	13	41	39	0.6	0.6
fabric, 8.0 oz/sq.yd	28	48	76	0.8	0.4
fabric, 9.3 oz/sq.yd	30	52	66	0.9	0.5
fabric, 9.6 oz/sq.yd	30	50	72	0.7	0.5
fabric, 10 oz/sq.yd	15	66	50	1.2	1.2
fabric, 10 oz/sq.yd	35	18	90	4.6	0.6
rug, 56 oz/sq.yd	25	324	464	1.5	0.5
Polyacetal, Celcon	125		6		NR
Nylon, 40% glass	100	487	41	3.63	9.20
unidentified		300	95	7.2	4.8
fabric, 13 oz/sq.yd	50	6	16	NR	15.0
carpet, foam backing, 86 oz	390	310	270		

(Continued)

Table 3.6. Smoke Density from Various Materials (National Bureau of Standards Test((Continued)

	Thickness mils	Maximum Specific Optical Density, Dm		Time to Ds = 16 minutes	
		nonflaming	flaming	nonflaming	flaming
Polyamide, aromatic,					
sheet, 17 oz/sq.yd	20	3	6	NR	NR
sheet, 64 oz/sq.yd	63	7	14	NR	NR
fabric, 4.4 oz/sq.yd	15	0	10	NR	NR
fabric, 6.1 oz/sq.yd	15	5	8	NR	NR
fabric, 9.9 oz/sq.yd	35	6	8	NR	NR
fabric, 11 oz/sq.yd	35	30	32	3.4	2.4
rug, 45 oz/sq.yd	300	65	51	4.8	3.2
carpet, 49 oz/sq.yd	220	175	105		
Polysulfone,					
sheet, 28 oz/sq.yd	31	1	28	NR	5.9
sheet, 54 oz/sq.yd	60	4	40	NR	2.6
P-1700	250	111	370	12.61	1.89
3M-360	125		55		8.3
Polycarbonate,					
Lexan 100	125		226		2.0
Lexan 113-T	250	48	324	10.85	1.95
Lexan 140	125		162		1.7
Lexan 2014	125		544		1.8
Lexan F6000	125	20	214		
Lexan MR4000	125	53	196		
Lexan F2000	125	38	334		
Lexan	62	3	275		
Lexan	62	1	165		
unidentified	125	12	174	NR	2.1
unidentified		21	324	27	2.0
unidentified, SE		44	660	14	1.6
Phenolic, Genal 4000	125	110	50	5	5.8
Genal 4200	125	137	55	5	5.8
Genal 4300	125	35	71	8	3.1
Genal 4301	125	20	41	20	5.6
Polyester,					
Paraplex P-43	125	780	780	2.66	0.59
Hetron 92	125	595		3.20	
glass reinforced	110	420	720		

(Continued)

Table 3.6. Smoke Density from Various Materials (National Bureau of Standards Test) (Continued)

	Thickness mils	Maximum Specific Optical Density, Dm nonflaming	flaming	Time to Ds = 16 minutes nonflaming	flaming
Urethane rigid foam, polyether, sucrose/ HT/PMDI,					
2.5 pcf	250	49	45	0.42	0.24
	500	70	80	0.46	0.20
	1000	111	71	0.44	0.24
	2000	221	113	0.35	0.22
3.2 pcf	250	62	47	0.52	0.60
	500	103	95	0.44	0.20
	1000	195	124	0.44	0.22
	2000	272	119	0.45	0.23
7.5 pcf	250	158	90	0.46	0.31
	500	272	112	0.53	0.25
	1000	446	134	0.52	0.24
	2000	499	183	0.57	0.33
16 pcf	250	363	161	0.72	0.38
	500	580	220	0.75	0.24
	1000	652	311	0.72	0.33
	2000	647	270	0.65	0.35
35 pcf	250	718	486	1.67	0.52
	500	658	710	1.61	0.52
	1000	652	529	1.46	0.41
	2000	710	789	1.32	0.66
62 pcf	250	780	664	2.22	0.69
	500	780	333	2.19	0.72
	1000	661	737	2.60	0.78
	2000	626	449	2.55	0.85
sucrose/MDI, FR, 2 pcf		170	500	0.30	0.13
TDI, FR, 1.8 pcf		290	285	0.44	0.17
TDI, 13 pcf		515	319	0.52	0.80
PAPI, FR, 2 pcf, CO_2 blown		119	196	0.36	0.21
PAPI, FR, 2 pcf		112	252	0.40	0.14
Urethane rigid foam, polyester,					
TDI, 2.5 pcf		161	70	0.43	0.20
TDI, 25 pcf		431	534	1.0	0.5
PAPI, 4 pcf		454	525	0.39	0.18

(Continued)

Table 3.6. Smoke Density from Various Materials (National Bureau of Standards Test) (Continued)

	Thickness mils	Maximum Specific Optical Density, Dm		Time to Ds = 16 minutes	
		nonflaming	flaming	nonflaming	flaming
Urethane rubber,					
TDI polyether		57	210	4.3	2.1
TDI polyester		131	230	4.0	1.5
Urethane foam, unidentified	500	156	20	0.5	0.5
Urethane foam, polyether	625	156	35	0.5	1.4
Urethane foam, polyester, 9.2 oz	210	77	58	0.6	1.9
Urethane foam, polyether, 35 oz	600	164	229	0.4	0.2
Urethane foam, polyether, FR	1000	318	41	0.7	0.6
Urethane foam, polyether, FR	1000	286	262	0.7	0.2
Urethane foam, polyether	1000	300	41	0.7	0.6
Red oak	250	395	76	4.1	8.0
Red oak	250	552			
Red oak	500	372	118	3.8	11.2
Red oak, flooring	750	505	300		
Red oak	780	660	117	7.1	7.8
White oak	250	420	107	3.5	6.6
White oak	500	409	56	4.0	11.8
Elm	50	150	35		
Elm	125	270			
Elm	185	390	65		
Elm	240	510			
Douglas fir	250	380	156	2.1	4.6
Douglas fir	500	438	110	3.0	6.8
Clear spruce	125	147	45	5.1	5.5
Clear spruce	250	275	115	4.3	4.2
Clear spruce	500	378	145	4.6	4.3
Clear spruce	750	421	310	4.8	4.8
Sitka spruce	500	263	130	4.2	7.7
White pine	250	325	155	2.3	2.7
Southern pine	500	431	156	3.0	7.8
Yellow birch	500	419	79	4.2	7.0

(Continued)

Table 3.6. Smoke Density from Various Materials (National Bureau of Standards Test) (Continued)

	Thickness Mils	Maximum Specific Optical Density, Dm		Time to Ds = 16 minutes	
		nonflaming	flaming	nonflaming	flaming
Black walnut	250	460	91	3.4	7.5
Western larch	500	323	111	3.8	7.4
Red lauan (Philippine mahogany)	500	374	96	2.5	5.4
Mahogany	500	320	109	3.5	5.2
Sugar maple	500	448	118	4.7	10.4
Eastern redcedar	500	372	76	2.9	4.6
Redwood	250	260	133	2.7	2.5
Redwood (Sequoia)	500	390	85	3.2	4.7
Marine plywood	250	285	62	2.7	5.4
Douglas fir interior plywood	250	350	96	3.4	5.3
Douglas fir exterior plywood					
uncoated	250	287	112	4.6	4.7
latex paint	254	258	108	3.8	5.6
alkyd paint	253	236	84	4.6	4.8
varnish	254	262	71	5.0	4.3
wallpaper, 5 coats	285	285	132	3.3	4.4
wall cloth	261	79	56	5.1	4.7
enameled wall covering	305	184	79	4.1	4.3
vinyl film	254	268	75	3.7	1.7
vinyl counter top	320	355	355	2.7	0.7
Douglas fir exterior plywood					
uncoated	375	404	14	1.9	4.8
primer, acrylic emulsion paint		336	4	2.0	NR
alkyd resin flat paint		195	21	3.1	7.0
phenolic spar varnish		422	14	1.6	6.3
shellac		488	12	1.9	3.8
alkyd enamel undercoat, alkyd resin titanium base flat paint		225	19	4.2	5.9
alkyd enamel undercoat, semigloss alkyd enamel		305	20	2.9	16.8

(Continued)

Table 3.6. Smoke Density from Various Materials (National Bureau of Standards Test) (Continued)

	Thickness mils	Maximum Specific Optical Density, Dm nonflaming	flaming	Time to Ds = 16 minutes nonflaming	flaming
Douglas fir exterior plywood					
alkyd enamel undercoat, gloss alkyd enamel		263	12	2.0	6.2
alkyd enamel undercoat, latex paint		255	23	2.1	15.1
oil base primer, linseed oil base paint		240	5	2.0	NR
linseed oil base paint		339	17	2.5	10.2
alkyd enamel undercoat, fire retardant latex paint		80	55	1.7	1.4
modified synthetic resin base fire retardant paint		44	69	1.7	2.4
modified epoxy fire retardant paint		195	137	0.9	0.3
sealer, modified polyurethane fire retardant varnish		361	117	0.8	0.2
Paperboard, homogeneous		359	68	1.3	4.2
Paperboard, laminated		169	60	2.0	10.9

(Concluded)

sion of particles, originating from combustion or sublimation.

Smoke is an important fire response characteristic because visibility is a factor in the ability of occupants to escape from a burning structure, and in the ability of firefighters to locate and suppress a fire.

Smoke evolution is most often expressed in terms of optical density because light and sight obscuration is the aspect of greatest concern.

Smoke evolution data on some materials are presented in Table 3.6.

SECTION 3.10. TOXIC GAS EVOLUTION

Toxic has been defined as poisonous, or destructive to body tissues

Table 3.7. Relative Toxicity of Off-Gases from Various Materials (PSC Procedure 1, or NASA-USF Procedure B)

Material	**Time to Death min**
Rigid Foams	
polymethacrylimide	11.45 ± 2.31
polybismaleimide	12.80 ± 1.04
polyisocyanurate, T1	21.22 ± 1.67
polyisocyanurate, T2	19.75 ± 0.79
polyisocyanurate, GM-43	21.14 ± 1.13
polyurethane, R1	23.92 ± 2.14
polyurethane, R2	23.74 ± 2.11
polyurethane, R3	25.69 ± 0.80
polyurethane, R4	24.93 ± 2.69
phenolic, GM-57	17.17 ± 0.83
polyvinyl chloride	24.40 ± 2.83
polystyrene	26.25 ± 0.64
Cellulosic Board	
hardboard	15.90 ± 2.62
medium density hardboard	17.87 ± 3.42
chipboard	18.23 ± 0.95
cellulose core board	20.58 ± 4.38
fiberboard soundstop	19.88 ± 5.04
fiberboard sheathing, asphalt impregnated	17.44 ± 2.05
Cellulose Insulation	
untreated	15.85 ± 1.15
16-37% boric acid	16.15 ± 0.97
14-36% 5-mol borax	19.11 ± 0.74
15-36% 46/54 BA/5MB	18.57 ± 2.45
Flexible Foams	
polyether urethane, F1	20.86 ± 2.71
polyether urethane, F2	19.91 ± 1.41
polyether urethane, F3	20.49 ± 0.63
polyether urethane, F4	19.41 ± 0.55
polyether urethane, F5	19.31 ± 0.80
polyether urethane, F6	20.10 ± 1.06
polyether urethane, F7	20.73 ± 1.12
polyether urethane, F8	22.38 ± 3.84

(Continued)

Table 3.7. Relative Toxicity of Off-Gases from Various Materials (PSC Procedure 1, or NASA-USF Procedure B)

Material	**Time to Death min**
Flexible Foams	
polyester urethane, 3000	18.91 ± 0.75
polyester urethane, 3003	18.54 ± 1.45
polyester urethane, 3013	16.17 ± 0.61
polyurethane, 1	18.55 ± 0.01
polyurethane, 2	21.99 ± 1.82
polyurethane, 3	22.73 ± 1.07
polyimide, 1	12.71 ± 0.02
polyimide, 2	14.76 ± 0.95
polyester, Pneumacel	18.97 ± 0.44
polyethylene, Microfoam	20.42 ± 0.26
polyethylene	20.93 ± 0.47
polychloroprene	24.51 ± 1.46
polychloroprene, ALS	26.84 ± 1.27
polychloroprene, LS	26.50 ± 2.52
polychloroprene, LS200	27.29 ± 0.42
polychloroprene, Vonar	25.83 ± 2.16
polychloroprene	24.29 ± 0.69
polyphosphazene, 1	23.98 ± 1.52
polyphosphazene, 2	24.11 ± 2.58
polysiloxane, 3-6548	25.00 ± 0.57
Fiber Battings	
cotton batting, untreated	9.66 ± 1.29
10% boric acid, padded on	14.58 ± 1.06
8% boric acid, vapor phase	11.20 ± 1.25
polyester batting	9.44 ± 0.76
13.3% acrylic resin	10.15 ± 0.64
20.6% acrylic resin	10.54 ± 1.28
Miscellaneous Cushioning	
feathers/down 75/25	7.26 ± 0.12
wool, washed	7.06 ± 1.18
rubberized hair	8.18 ± 0.50
rubberized hair, FR	7.62 ± 0.83
excelsior	15.82 ± 0.11
sisal	12.59 ± 3.41
kapok	16.15 ± 0.47

(Continued)

Table 3.7 Relative Toxicity of Off-Gases from Various Materials (PSC Procedure 1, or NASA-USF Procedure B) (Continued)

Material	Time to Death min
Food and Food Packaging	
mashed potatoes	10.90 ± 1.10
soft drink mix	13.30 ± 0.85
sugar, pure cane, granulated	14.68 ± 0.38
black pepper, pure ground	15.09 ± 1.21
crackers, saltine	15.14 ± 1.40
gelatin dessert	15.46 ± 0.16
wafers, vanilla	16.86 ± 1.18
cooking oil	17.45 ± 0.33
oats, rolled	17.94 ± 0.34
margarine	18.25 ± 0.32
freeze-dried coffee	18.63 ± 1.09
chocolate flavoring	18.85 ± 0.18
non-dairy cramer	19.34 ± 1.15
flour, white	19.53 ± 2.31
spaghetti, enriched	19.65 ± 0.97
rice, enriched, long grain	20.19 ± 0.95
cocoa mix	20.92 ± 1.48
dry milk, nonfat	21.05 ± 0.24
egg yolk, hard cooked	24.58 ± 2.14
egg white, hard cooked	27.24 ± 1.10
paper towels	15.46 ± 1.31
egg carton, polystyrene foam	28.63 ± 0.52

(Concluded)

and organs or interfering with body functions. The toxic gases evolved from materials involved in fires may lead to incapacitation and death, and are the principal threat to life safety when the thermal threat is minor or insignificant.

Organic materials when decomposing and burning can evolve a variety of toxic gases, the most common of which is carbon monoxide. In addition, nitrogen-containing materials may evolve hydrogen cyanide and nitrogen oxides, chlorine-containing materials may evolve hydrogen chloride, and sulfur-containing materials may evolve sulfur oxides and sulfides.

Table 3.8. Relative Toxicity of Off-Gases from Materials by Generic Type (PSC Procedure 1, or NASA-USF Procedure B)

Material	Number of Samples	Time to Death min
Plastics and Wood		
Polyether sulfone	3	12.30 ± 2.08
Polyphenylene sulfide	4	13.21 ± 3.80
Polyaryl sulfone	2	13.48 ± 3.17
Polyimide flexible foam	2	13.74 ± 1.45
Wood	12	14.03 ± 1.48
Polyamide (nylon)	3	14.36 ± 1.71
Polyphenyl sulfone	1	15.46
Polyurethane rigid foam	7	15.49 ± 4.06
Polymethyl methacrylate (PMMA)	1	15.58
Polyvinylidene fluoride	1	15.86
Cellulosic board	8	16.57 ± 3.54
Polyvinyl chloride (PVC)	2	16.60 ± 0.33
Acrylonitrile/butadiene/styrene (ABS)	3	17.13 ± 2.45
Phenolic	1	17.17
Polyetherimide	1	17.40
Polyethylene, including foam	5	17.31 ± 3.73
Acrylonitrile rubber (NBR)	3	19.13 ± 2.89
Polyphenylene oxide, modified	1	19.96
Polyurethane flexible foam	14	20.01 ± 1.74
Bisphenol A polycarbonate	3	20.40 ± 3.77
Polyvinyl fluoride	1	20.50
Ethylene/propylene/diene (EPDM)	2	20.69 ± 0.04
Chlorosulfonated polyethylene	2	20.88 ± 2.07
Polyisocyanurate rigid foam	2	21.68 ± 1.38
Polyisoprene (natural rubber)	1	22.13
Chlorinated polyvinyl chloride (CPVC)	2	22.25 ± 0.69
Polystyrene	2	23.10 ± 4.33
Polyphosphazene flexible foam	2	24.05 ± 0.09
Styrene-butadiene rubber (SBR)	1	24.11
Polychloroprene flexible foam	6	25.88 ± 1.24
Chlorinated polyethylene	2	26.08 ± 1.80

(Continued)

Table 3.8. Relative Toxicity of Off-Gases from Materials by Generic Type (PSC Procedure 1, or NASA-USF Procedure B) (Continued)

Material	**Number of Samples**	**Time to Death min**
Fibers and Fabrics		
Wool, 100%	4	7.65 ± 1.29
Wool, 85–90%/nylon	4	8.87 ± 1.01
Nylon, 100%	9	16.78 ± 3.49
Silk, 100%	2	9.18 ± 0.35
Silk, 70%/rayon, 30%	2	12.33 ± 0.58
Rayon, 100%	10	15.40 ± 2.41
Polyester, 100%	3	10.70 ± 2.25
Polyester, 65–87%/cotton	3	10.45 ± 0.26
Cotton, 59–70%/polyester	2	15.66 ± 1.17
Cotton, 100%	10	13.08 ± 2.14
Cotton, 100%	10	13.08 ± 2.14
Cotton, 82–86%/rayon	2	12.00 ± 0.16
Cotton, 52–75%/rayon	8	14.53 ± 2.06
Rayon, 54–75%/cotton	18	12.70 ± 2.49
Rayon, 100%	10	15.40 ± 2.41
Rayon, 100%	10	15.40 ± 2.41
Rayon, 56–73%/polypropylene	2	14.08 ± 2.16
Polypropylene, 100%	4	16.64 ± 2.76
Rayon, 100%	10	15.40 ± 2.41
Rayon, 52–72%/nylon	3	15.60 ± 3.86
Nylon, 57–62%/rayon	2	15.62 ± 0.15
Nylon, 100%	9	16.78 ± 3.49
Cotton, 100%	10	13.08 ± 2.14
Cotton, 100%, FR	6	14.19 ± 3.64

(Concluded)

Because identification and analysis of all toxicants is difficult, and the combined effect of a group of toxicants of varying concentrations and toxicities cannot be accurately predicted with the present state of the art, laboratory animals are used to integrate all the toxic effects of a specific mixture of evolved gases.

CHAPTER 4

Flammability Tests

Because expressing information as numbers facilitates comparison, the behavior of plastic materials when exposed to fire is generally expressed as numbers obtained through flammability tests. It is hoped that these tests will provide some correlation with actual fire exposure. Most tests are relatively small in scale, because experimental materials are available in relatively small quantities, because it is prohibitively costly to burn complete structures repeatedly, and because small-scale tests are more easily replicated than large-scale fires. Many tests have an inherent deficiency in that they fail to reproduce the massive effect of heat present in a large-scale fire, and therefore give results that can be misleading if applied in the wrong context.

SECTION 4.1. TYPES OF FLAMMABILITY TESTS

Flammability tests can be classified according to the fire response characteristics which they are designed to measure:

1. Tests for smolder susceptibility, such as the tests developed by the California Bureau of Home Furnishings.
2. Tests for ignitability, such as the ASTM D 1929 test for ignition temperatures.
3. Tests for flash-fire propensity, such as the test developed by the Douglas Aircraft Company.
4. Tests for flame spread, such as the ASTM E 84 25-foot tunnel test.

5. Tests for heat release, such as the Ohio State University release rate test.

6. Tests for fire endurance, such as the ASTM E 119 and UL 181 tests.

7. Tests for ease of extinguishment, such as the ASTM D 2863 test for oxygen index.

8. Tests for smoke evolution, such as the ASTM E 662 and D 2843 tests.

9. Tests for toxic gas evolution, such as the DIN 53 436 test.

Tests can be divided into two general groups: those which are primarily research tests, and those which are primarily acceptance tests. The first group includes those test methods which are standard primarily because they are recognized as desirable methods for evaluating the flammability characteristics of materials on the basis of technical soundness and scientific value. The second group includes those test methods which are standard primarily because they are a requirement for doing business. Most of the acceptance tests have, at some time or other, been used as research tests, and hopefully the best research tests will eventually become acceptance tests.

Flammability tests can be divided into low-severity tests, which provide exposure to relatively small amounts of heat for relatively short periods of time, simulating small accidental fires, and high-severity tests which simulate the large-scale fire with massive heat exposure.

Flammability tests can be divided into those which are intended for individual materials, and those designated for complete systems. Most tests for fire endurance and many tests for surface flame spread are designed for complete systems. Most tests for ease of ignition and many tests for surface flame spread and smoke evolution are intended for individual materials.

Flammability tests for plastic materials can be divided into four groups on the basis of origin:

1. Tests originally designed for plastics. These tend to be low-severity tests because the initial uses for plastics were in small articles and in parts for electrical appliances.

2. Tests originally designed for materials of construction. These tend to be high-severity tests, and have become pertinent to plastics because of their increasing use in construction.

3. Tests originally designed for measuring flammability hazard in a particular application or under a specific type of exposure, such as hot-bolt and arc-ignition tests, and tests for pipe insulation.

4. Tests originally designed for fabrics and textiles, and pertinent to plastics because of their increasing use in home furnishings and vehicle interiors.

SECTION 4.2. TESTS FOR SMOLDER SUSCEPTIBILITY

Smolder susceptibility may be defined as the tendency of a material to support smoldering combustion within itself. This characteristic provides a measure of fire hazard in that a material undergoing smoldering combustion may generate increasing amounts of heat and, under certain conditions, bring itself or an adjacent material to the point of ignition.

Test methods for smolder susceptibility have been developed by the National Bureau of Standards (4.201) and by the California Bureau of Home Furnishings (4.202, 4.203). Both methods provide for the use of a small-scale mock-up 203 mm (8 in) wide primarily for comparing fabrics, and a larger-scale prototype 450 by 550 mm (18 by 22 in) simulating complete assemblies. The ignition source is a king-size non-filter cigarette covered with one layer of laundered 100% cotton bed sheeting material. Some smolder susceptibility data from the California Bureau of Home Furnishings (4.203) are presented in Table 3.1.

A review of the data shows that smolder susceptibility is not a function of the fabric alone, or of the substrate alone, but of the specific fabric/substrate combination. The fiberglass board substrate appears to be a well-intentioned pseudo-scientific selection which provides neither relevance to real life nor impartial safety factors.

The 100% cellulosic fabrics exhibited the greatest susceptibility to smoldering, but this susceptibility was greatly reduced when the substrate was resinated polyester batting or neoprene foam interliner. This smolder susceptibility was also reduced when the substrate was non-resinated polyester batting or certain polyurethane flexible foams (three out of four samples in the study). The 100% thermoplastic fabrics exhibited no smolder susceptibility except when untreated cotton batting was the substrate.

Smoldering combustion generally requires a porous fuel, and charring appears to be a prerequisite for smoldering. Cellulosic fabrics such as cotton and rayon are, like wood, prone to char and smolder. The char formation promoted by some fire retardants also encourages smoldering. Thermoplastic fabrics such as nylon and polyolefin inhibit smoldering, by not forming char and by melting down to help seal off areas of char to air diffusion and spread of smoldering. Wool and wool blend fabrics, which both melt and char, tend to exhibit enough melting to inhibit smoldering combustion. The marked differences between these types of fabrics are brought out in smolder susceptibility data using the USF smolder susceptibility test method, which uses a California-type test stand and polyurethane flexible foam substrates of the appropriate densities for seat back and bottom cushions, fire-retarded to meet FAA and California flame-resistance requirements (4.204); these data are presented in Table 4.1.

Table 4.1. Smolder Susceptibility of Materials as Measured by the USF Smolder Susceptibility Test Method

Upholstery Fabric Material	Smolder Time minutes	Foam Weight Loss, grams
100% polyolefin (furniture)	25.1 ± 1.7	0
100% wool (aircraft)	28.2 ± 1.7	0
100% wool (aircraft)	29.4 ± 3.9	0
83% wool, 17% nylon (aircraft)	29.6 ± 0.8	0
49% wool, 51% PVC (aircraft)	31.0 ± 3.7	0
100% cotton (furniture)	63.1 ± 8.0	2.8 ± 1.5
100% rayon (aircraft)	67.6 ± 1.6	1.4 ± 0.7

SECTION 4.3. TESTS FOR FLASH-FIRE PROPENSITY

Flash-fire propensity may be defined as the tendency of a material to produce a fire that spreads with extreme rapidity. This characteristic provides a measure of fire hazard in that a material with high flash-fire propensity is capable to contributing to sudden development of a rapidly spreading fire.

The basic approach to laboratory evaluation of flash-fire propensity involves propagation through fuel-air mixtures in a tube. Most laboratory methods use upward propagation. The flammable limits are wider for upward propagation than for horizontal or downward propagation, and are favored for safety considerations.

The National Bureau of Standards apparatus (4.401) employs a tube 50 mm in diameter and 500 mm long, into which pyrolysis gases are introduced and ignited by an electrical spark. A sample of known weight is introduced into a furnace at 500°C to produce the gas mixture which is recirculated to improve uniformity. Some NBS flash-fire data (4.401) are presented in Table 4.2.

Table 4.2. Flash-Fire Propensity of Materials as Measured by the NBS Flash-Fire Test

Material	Minimum Mass grams/liter
Polyphenylene oxide, modified	0.23
Polyester urethane flexible foam, 1	0.41
Wool carpet, polyester backing, FR latex backcoating	0.55
Wool carpet, polyester backing, FR latex backcoating, polyurethane foam pad, nylon scrim	0.55
Wool fabric, FR	0.62
Polyvinyl chloride flexible foam	0.67
Polyester urethane flexible foam, 2	0.72
Polyester-fiberglass sheet	0.84
Polycarbonate, molded	1.20
Polycarbonate, transparency	1.25
Epoxy-fiberglass faces, aramid honeycomb	2.91
PVF/PVC finish, phenolic-fiberglass faces, epoxy adhesive, aramid honeycomb	3.30
PVF perforated finish, aramid-epoxy face, aramid honeycomb core, fiberglass-epoxy face	4.30
FR coating, cotton/rayon fabric	greater than 1.37
PVF finish, aramid-phenolic finish	1.41
Wool carpet, phenolic/epoxy-fiberglass face, epoxy adhesive, aramid honeycomb, epoxy-fiberglass face	1.70
Fiberglass batting, melamine binder	1.83
Epoxy-fiberglass sheet	2.04
Aramid fabric	2.09
PVF finish, fiberglass screen, aramid honeycomb core with fiberglass batting, phenolic/epoxy-fiberglass face	2.35
Fiberglass batt, silicone-treated, phenolic-impregnated	2.35
PVF finish, epoxy-fiberglass faces, aramid honeycomb	2.77
PVC finish, phenolic/epoxy-fiberglass face, epoxy adhesive, aramid honeycomb, epoxy-fiberglass face	3.67
Polyether urethane foams	
ungrafted, no FR	0.1917
polyacrylonitrile grafted, no FR	0.2385
polyacrylonitrile grafted, FR (Br,P)	0.3674
ungrafted, FR (Br,P)	0.3888

The Douglas Aircraft Company apparatus (4.402) uses a combustion tube of similar design, with an improved pyrolysis method and without recirculation. A 0.50 g sample is pyrolyzed at a heating rate of approximately 500°C/min to produce the pyrolysis gases. Some Douglas flash-fire data (4.402) are presented in Table 4.3.

Table 4.3. Flash-Fire Propensity of Materials as Measured by the Douglas Flash-Fire Test

Material	**Time to Flash Fire, min**	**Sample Temp. at Flash, °C.**
Nylon fabric	1.2	570
Polyurethane flexible foam	1.36	600
Polyurethane-coated nylon fabric	1.36	450
Wool/nylon fabric	1.56	275
Polychloroprene flexible foam, 1	1.6	740
Phenolic fabric	1.6	850
Phenolic/aramid fabric, 1	1.72	750
Polyamide-imide fabric	2.86	910
Silicone flexible foam, 1	2.96	930
Silicone flexible foam, 2	3.0	825
Phenolic/aramid fabric, 2	3.1	940
Phenolic/aramid fabric, 3	3.28	810
Polybenzimidazole fabric	4.18	940
Polyphosphazene flexible foam	none	
Silicone flexible foam, 3	none	
Polychloroprene flexible foam (4 samples)	none	
Polyamide-imide/wool fabric	none	
Aramid fabric (2 samples)	none	

The University of San Francisco flash-fire screening test method (4.403) uses a vertical combustion tube 45 mm in diameter and 660 mm high, into which pyrolysis gases from a horizontal tube furnace are introduced, permitted to mix with air, and ignited by the hot pyrolyzing surfaces. A 0.10 g sample is introduced into the furnace at 800°C to produce the pyrolysis gases. The height of the flash fire and the time to flash fire are recorded. Some USF flash-fire data (4.403) are presented in Table 3.3.

Polyurethane flexible foams were among the materials exhibiting the greatest flash-fire propensity in all three tests, and polychloroprene flexible foams were among the materials exhibiting the least flash-fire propensity in the Douglas and USF tests. The materials that appeared to be the least prone to flash fires included polyvinyl chloride, polyphenylene oxide and sulfide, polyether and polyaryl sulfone, and polychloroprene.

Flash fires developed most rapidly with cellulosic fabrics and cushioning materials, wood, and polyurethane flexible foam. The largest flash fires were observed to occur with wood, polyolefin plastics and fabrics, nylon plastics and fabrics, and polyurethane flexible foam. For these materials which are large-volume products and not easily modified, careful design of the application and limitation of the amount exposed seem advisable.

The NBS and Douglas flash-fire test data presented were derived from studies directed toward aircraft interior materials, and therefore do not cover many materials used in residential, commercial, and industrial environments. Polyurethane flexible foam was the only generic material given significant attention in the NBS, Douglas, and USF work. Differences between the NBS and USF data regarding the minimum sample-to-volume ratio needed to produce a flash fire with polyurethane foam appear to be due to the difference in pyrolysis temperatures of 500°C and 800°C, respectively (4.404).

Dust explosions are similar to flash fires in that they can occur only within certain ranges of combustible concentrations and are characterized by extremely rapid movement of a flame front. Dust explosions are influenced by additional factors such as particle size and dispersion.

The dust explosibility apparatus which has been most extensively used is the Hartmann apparatus developed by the U.S. Bureau of Mines (4.405). The basic Hartmann apparatus is a vertical cylinder of polymethylmethacrylate, with a height of 12 in and a volume of 75 in^3, mounted on a precision-machined base which acts as the sample holder. Several test procedures have been developed. The ignition sensitivity is calculated from the minimum ignition temperature, the minimum ignition energy, and the minimum explosive concentration. The explosion severity is calculated from the maximum explosion pressure and the maximum rate of pressure rise. The explosibility index is the product of the ignition sensitivity and the explosion severity. Some dust explosibility data are presented in Table 4.4.

Table 4.4. Explosibility of Dusts of Various Polymers

Polymer	Minimum Ignition Temp., °C.		Minimum Explosive Conc. oz/cu.ft.	Minimum Ignition Energy joule	Ignition Sensitivity	Explosion Severity	Explosibility Index
	cloud	layer					
Polyethylene							
type D	450		0.025	0.080	2.2	1.1	2.4
high pressure, sample 1	450	380	0.020	0.030	7.5	1.4	>10
high pressure, sample 2	410					0.8	
low pressure, sample 1	420		0.020	0.060	4.0	1.0	4.0
low pressure, sample 2	450		0.020	0.010	22.4	2.3	>10
low pressure, melt index 0.4	430					1.4	
low pressure, melt index 6.0	420					2.2	
wax, low molecular weight	400		0.020	0.035	7.2	0.8	5.8
Polypropylene							
molecular weight 1.8 million	460		0.035	0.400	0.3	0.2	0.1
molecular weight 1.1 million	460		0.030	0.025	5.8	0.4	2.3
molecular weight 0.6 million	460		0.055	0.400	0.2	0.1	<0.1
linear	420		0.020	0.030	8.0	1.0	8.0
antioxidant	420		0.020	0.030	8.0	2.0	>10
0.3–0.4% antioxidant	420					1.7	
Polyvinyl acetate	550		0.040	0.160	0.6	0.4	0.2
Polyvinyl acetate alcohol	520	440	0.035	0.120	0.9	1.2	1.1
Polyvinyl butyral	390		0.020	0.010	25.8	0.9	>10

(Continued)

Table 4.4. Explosibility of Dusts of Various Polymers

Polymer	Minimum Ignition Temp., °C. cloud	Minimum Ignition Temp., °C. layer	Minimum Explosive Conc. oz/cu.ft.	Minimum Ignition Energy joule	Ignition Sensitivity	Explosion Severity	Explosibility Index
Polyvinyl chloride							
sample 1	730	290	>2.00	>8.32	< 0.1		< 0.1
sample 2, fine	660	400	>2.00	>8.32	< 0.1	<0.1	< 0.1
sample 3, coarse	690	480	>2.00	>8.32	< 0.1	<0.1	< 0.1
sample 4, powdered	680	400	>2.00	>8.32	< 0.1	<0.1	< 0.1
copolymer	720	500	>2.00	>8.32	< 0.1	<0.1	< 0.1
binder for fiber batting	670		>2.00	>8.32	< 0.1	<0.1	< 0.1
Vinyl chloride-vinyl acetate copolymer							
sample 1	690		>2.00	>8.32	< 0.1		< 0.1
sample 2	750		>2.00	>8.32	< 0.1		< 0.1
sample 3	710		>2.00	>8.32	<0.1		< 0.1
molding compound, mineral filler	690		>2.00	>8.32	< 0.1		< 0.1
Vinyl chloride-acrylonitrile							
60/40 copolymer, water emuls. prod.	570	470	0.045	0.025	3.1	0.6	1.9
33/67 copolymer, water emuls. prod.	530	470	0.035	0.015	7.2	2.0	>10
Vinyl chloride-polyoctyl acrylate							
79/21 copolymer	500	430	0.100	0.960	< 0.1	<0.1	< 0.1
Vinyl chloride-diisopropyl fumarate							
70/30 copolymer	580		0.060	0.060	1.0	0.9	0.9
Polyvinyl chloride-dioctyl phthalate							
67/33 mixture	320		0.035	0.050	3.6	0.8	2.9
Polyvinyl chloride-Hycar rubber							
copolymer, sample 1	490		0.025	0.030	5.5	1.7	9.4
copolymer, sample 2, more resin	550	460	0.070	0.060	0.9	0.3	0.3

(Continued)

Table 4.4. Explosibility of Dusts of Various Polymers (Continued)

Polymer	Minimum Ignition Temp., °C		Minimum Explosive Conc. oz/cu.ft.	Minimum Ignition Energy joule	Ignition Sensitivity	Explosion Severity	Explosibility Index
	cloud	layer					
Vinyl-vinylidene chloride copolymer							
mainly vinyl	780	450	>2.00	>8.32	< 0.1		< 0.1
mainly vinylidene	>1000	420	>2.00	>8.32	< 0.1		< 0.1
Vinylidene copolymer, 10% plasticizer	830	390	>2.00	>8.32	< 0.1		< 0.1
Vinylidene chloride polymer mold. cpd.	900		>2.00	>8.32	< 0.1		< 0.1
Vinyl multipolymer							
monomeric vinylidene cyanide	500	510	0.030	0.015	8.9	3.0	>10
Vinyl toluene-acrylonitrile-butadiene							
58/19/23 copolymer, sample 1	530		0.020	0.020	9.5	1.6	>10
58/19/23 copolymer, sample 2	530		0.020	0.020	9.5	2.2	>10
Polyvinyl toluene, sulfonated	540	330	1.000	2.880	< 0.1		< 0.1
Polyvinyl benzyl trimethyl ammonium							
chloride, flake	420	240	0.045	0.140	0.8	0.3	0.2
yellow, divinyl benzene	410	220	0.035	0.100	1.4	0.7	1.0
Polystyrene, clear	490		0.020	0.120	1.7	0.5	0.9
special grind	500		0.020	0.040	5.0		
molding compound	560		0.015	0.040	6.0	2.0	>10
beads	500	470	0.025	0.060	2.7	1.5	4.1
latex, spray dried	500	500	0.020	0.015	13.4	3.3	>10
Styrene-hydrocarbon							
monomer polymer							
85/15 copolymer, sample 1	460	450	0.020	0.035	6.3	2.1	>10
85/15 copolymer, sample 2	460	470					

(Continued)

Table 4.4. Explosibility of Dusts of Various Polymers (Continued)

Polymer	Minimum Ignition Temp., °C. cloud	Minimum Ignition Temp., °C. layer	Minimum Explosive Conc. oz/cu.ft.	Minimum Ignition Energy joule	Ignition Sensitivity	Explosion Severity	Explosibility Index
Styrene-acrylonitrile 70/30 copolymer	500		0.035	0.030	3.8	0.5	1.9
Polystyrene-Buna N rubber							
coprecipitate	510	500	0.020	0.080	2.5	2.3	5.8
Styrene-butadiene latex copolymer							
less than 4% zinc stearate blend	470		0.030	0.060	2.4	0.6	1.4
over 75% styrene, alum coagulated	440		0.025	0.025	7.3	1.7	>10
Methyl methacrylate polymer							
sample 1	480		0.030	0.020	7.0	0.9	6.3
sample 2, molding compound, fines	440		0.020	0.015	15.3	1.0	>10
sample 3, molding compound, fines	440		0.030	0.020	7.6	0.8	6.1
Methyl methacrylate-ethyl acrylate							
copolymer, sample 1	480		0.030	0.010	14.0	2.7	>10
sample 2, spray dried	500		0.035	0.025	4.6	2.0	9.2
Methyl methacryalte-ethyl acrylate-							
styrene copolymer	440		0.025	0.020	9.2	1.7	>10
Methyl methacrylate-styrene-butadiene-							
acrylonitrile copolymer	480		0.025	0.020	8.4	1.4	>10
Methyl methacrylate-styrene-butadiene-							
ethyl acrylate copolymer	480		0.025	0.025	6.7	1.5	>10
Methacrylic acid polymer, modified	450	290	0.045	0.100	1.0	0.6	0.6
Isobutyl methacrylate polymer	500	280	0.020	0.040	5.0	1.0	5.0
Acrylamide polymer	410	240	0.040	0.030	4.1	0.6	2.5

(Continued)

Table 4.4. Explosibility of Dusts of Various Polymers (Continued)

Polymer	Minimum Ignition Temp., °C.		Minimum Explosive Conc.	Minimum Ignition Energy	Ignition Sensitivity	Explosion Severity	Explosibility Index
	cloud	layer	oz/cu.ft.	joule			
Acrylamide-vinyl benzyl trimethyl ammonium chloride copolymer	810	500	1.000	8.000	< 0.1		< 0.1
Acrylonitrile polymer	500	460	0.025	0.020	8.1	2.3	>10
Acrylonitrile-vinyl pyridine copolymer	510	240	0.020	0.025	7.9	2.4	>10
Acrylonitrile-vinyl chloride-vinylidene chloride copolymer 70/20/10	650	210	0.035	0.015	5.9	3.0	>10
Cellulose acetate, sample 1	420		0.040	0.015	8.0	1.6	>10
sample 2	420		0.035	0.030	4.6	0.9	4.1
sample 3	420	420	0.040	0.045	2.7		
sample 4	420	420	0.050	0.045	2.1		
sample 5	470	400	0.045	0.025	3.8	3.0	>10
sample 6	460	430	0.045	0.040	2.4	3.2	7.7
sample 7	400	400	0.040	0.050	2.5	2.2	5.5
sample 8	460	430	0.040	0.035	3.1	2.1	6.5
sample 9	480	380	0.040	0.045	2.3	2.9	6.7
sample 10	420	400	0.040	0.050	2.4	2.4	5.8
sample 11	430		0.045	0.030	3.5	3.2	>10
sample 12	440	340	0.055	0.020	4.2	3.1	>10
sample 13	430		0.035			1.5	
sample 14, 5–10 micron	450	390	0.035	0.020	6.4	3.7	>10
sample 15, molding compound	410		0.035	0.040	3.5	0.9	3.2
sample 16, 54.5% acetyl, spinning	450		0.040	0.080	1.4	2.5	3.5
sample 17, 53.0% acetyl, molding	470		0.035	0.060	2.0	2.3	4.6

(Continued)

Table 4.4. Explosibility of Dusts of Various Polymers (Continued)

Polymer	Minimum Ignition Temp., °C. cloud	Minimum Ignition Temp., °C. layer	Minimum Explosive Conc., oz/cu.ft.	Minimum Ignition Energy joule	Ignition Sensitivity	Explosion Severity	Explosibility Index
Cellulose triacetate, sample 1	430		0.035	0.030	4.5	1.2	5.4
sample 2	430		0.040	0.030	3.9	1.9	7.4
Cellulose acetate butyrate, sample 1	410		0.035	0.030	4.7	1.2	5.6
sample 2	440		0.035			1.5	
sample 3, molding compound	370		0.025	0.030	7.3	1.1	8.0
Cellulose acetate butyrate-cellulose acetate mixture	430		0.030			2.8	
Cellulose propionate, 0.3% free OH	460		0.025	0.060	2.9	2.6	7.5
Cellulose tripropionate, 0% free OH	460		0.025	0.045	3.9	1.8	7.0
Ethyl cellulose, sample 1			0.020				
sample 2, 5–10 micron	370	350	0.025	0.020	21.8	3.4	>10
sample 3, no fill, or plast.	340	330	0.025	0.015	15.8	3.6	>10
sample 4, molding compound	320		0.025	0.010	25.2	3.2	>10
Methyl cellulose, no fill. or plast.	360	340	0.030	0.020	9.3	3.1	>10
Carboxy methyl cellulose, sample 1	350	290				0.2	
low visc., 0.3–0.4% subs.	450	290	0.165	0.180	0.2	0.2	< 0.1
low visc., 0.3–0.4% subs., acid prod.	460	310	0.060	0.140	0.5	2.7	1.4
med visc., 0.84% subs.	370	260	0.150	0.560	0.1	0.5	< 0.1
0.65–0.95% subs.	360		0.400	1.920	< 0.1	0.2	< 0.1
0.65–0.95% subs.	330		0.350	0.800	< 0.1	0.2	< 0.1
0.2–0.3% subs.	400		0.300	1.920	< 0.1	<0.1	< 0.1
0.98% subs., 56.4% active agent	400	380	0.340	>8.32	< 0.1	0.2	< 0.1

(Continued)

Table 4.4. Explosibility of Dusts of Various Polymers (Continued)

Polymer	Minimum Ignition Temp., °C. cloud	layer	Minimum Explosive Conc., oz/cu.ft.	Minimum Ignition Energy joule	Ignition Sensitivity	Explosion Severity	Explosibility Index
Carboxy methyl hydroxyethyl cellulose							
sample 1	400	330	0.250	1.280	< 0.1	0.1	< 0.1
sample 2, 0.65–0.85% subs.	380		0.200	0.960	< 0.1	0.3	< 0.1
Hydroxyethyl cellulose-monosodium phosphate, sizing compound	390	340	0.070	0.035	2.1	0.8	1.7
Acetal, linear (polyformaldehyde)	440		0.035	0.020	6.5	1.9	>10
Nylon (polyhexamethylene adipamide)							
sample 1	500	430	0.030	0.020	6.7	1.8	>10
sample 2	510		0.050	0.030	2.6	3.3	8.6
sample 3, chemically precipitated	540		0.035	0.030	3.6	1.1	4.0
Chlorinated polyether alcohol	460		0.045	0.160	0.6	0.3	0.2
Ethylene oxide polymer	350		0.030	0.030	6.4	0.9	5.8
Polycarbonate	710		0.025	0.025	4.5	1.9	8.6
Phenol formaldehyde, sample 1	580		0.025	0.015	9.3	1.4	>10
sample 2	670		0.035	0.025	3.4		
sample 3	730		0.035	0.080	1.0	1.9	1.9
sample 4	700		0.025	0.035	3.3	1.7	5.6
sample 5	580		0.025	0.020	6.9	3.9	>10
sample 6, powdered	630		0.175	>8.32	< 0.1	0.1	< 0.1
sample 7, spray dried	580		0.175	3.840	< 0.1	0.1	< 0.1
sample 8, spray dried	660	320	0.200	6.000	< 0.1	0.1	< 0.1
sample 9, alkaline	620		0.040	0.030	2.7		

(Continued)

Table 4.4. Explosibility of Dusts of Various Polymers (Continued)

Polymer	Minimum Ignition Temp., °C. cloud	Minimum Ignition Temp., °C. layer	Minimum Explosive Conc., oz/cu.ft.	Minimum Ignition Energy joule	Ignition Sensitivity	Explosion Severity	Explosibility Index
Phenol formaldehyde,							
sample 10, 1-step	640		0.040	0.010	7.9	5.3	>10
sample 11, 2-step	580		0.025	0.010	13.9	4.0	>10
sample 12, 2-step	580		0.030	0.015	7.7	4.1	>10
sample 13, 2-step	590		0.035	0.030	3.2		
sample 14, novalac, angular part.	620		0.030	0.020	5.4	2.5	>10
sample 15, novalac, spherical part.	650		0.035	0.030	2.9	0.8	2.3
sample 16, hollow spherical part.	500	190	0.250	>8.32	< 0.1	0.2	< 0.1
sample 17, hollow spherical part.	490	180	0.250	>8.32	< 0.1	0.3	< 0.1
sample 18, infusible, insoluble	500	210	0.200	1.300	< 0.1	<0.1	< 0.1
sample 19, 1.5% zinc stearate and 1% oxalic acid	550	300	0.025	0.010	14.6	2.7	>10
sample 20, fiber batting binder	600		0.025	0.025	5.4	3.5	>10
sample 21, semi resinous	460		0.235	>8.32	< 0.1	<0.1	< 0.1
sample 22, glass and fibers	590						
sample 23, more fibers than above	540						
sample 24, 20% cellulosic extender	500		0.120	3.840	< 0.1	0.1	< 0.1
Phenol formaldehyde molding compound							
sample 1, cotton flock filler	490		0.030	0.010	13.7	5.3	>10
sample 2, wood flour filler	500		0.030	0.015	8.9	4.7	>10

(Continued)

Table 4.4. Explosibility of Dusts of Various Polymers (Continued)

Polymer	Minimum Ignition Temp., °C		Minimum Explosive Conc., oz/cu.ft.	Minimum Ignition Energy joule	Ignition Sensitivity	Explosion Severity	Explosibility Index
	cloud	layer					
Phenol formaldehyde, C stage, modified by hydrophilic groups,							
sample 1	440						
sample 2	590						
sample 3	500						
Phenol formaldehyde, amine modified	510		0.070			0.1	
Phenol formaldehyde, polyalkylene polyamine modified	420	290	0.020	0.015	16.0	2.8	10
Phenol anhydro formaldehyde anilin, 2-step	570		0.035	0.010	10.1	5.1	10
Phenol formaldehyde derivative, calcium,							
sample 1, spray dried	460	480	0.030	0.025	5.8	1.0	5.8
sample 2, spray and drum dried	460	480	0.025	0.020	8.8	3.6	>10
Phenol formaldehyde, sulfonated	640	430	>2.00	>8.32	< 0.1		< 0.1
Melamine formaldehyde, no plasticizer	810		0.085	0.320	0.1	0.2	< 0.1
Melamine formaldehyde, plasticizer	790		0.065	0.050	0.8	0.9	0.7
Urea formaldehyde,							
sample 1	530		0.135	1.280	0.1	0.1	0.1
sample 2, glue	510		0.070	0.640	0.1	0.4	0.1
sample 3, glue	510		0.075	0.960	0.1	0.8	0.1
sample 4, glue	470		0.070	0.080	0.8	0.6	0.5

(Continued)

Table 4.4. Explosibility of Dusts of Various Polymers (Continued)

Polymer	Minimum Ignition Temp., °C. cloud	layer	Minimum Explosive Conc., oz/cu.ft.	Minimum Ignition Energy joule	Ignition Sensitivity	Explosion Severity	Explosibility Index
Urea formaldehyde molding compound							
sample 1, granular	480		0.165	0.080	0.3	0.2	0.1
sample 2	530		0.090	0.080	0.5	1.1	0.6
sample 3, grade I, medium fine	450		0.075	0.160	0.4	0.9	0.4
sample 4, grade II, fine	460		0.085	0.080	0.6	1.7	1.0
sample 5, wood flour filler	490	530	0.075	0.160	0.3	0.9	0.3
Urea formaldehyde-phenol formaldehyde molding compound, wood flour filler	530	240	0.085	0.120	0.4	0.6	0.2
Epoxy, one part anhydride type, 1% cat.	530		0.030	0.035	3.6	2.0	7.2
Epoxy, no cat., modif., no addit.	540		0.020	0.015	12.4	2.7	>10
Epoxy-bisphenol A mixture	510		0.030	0.035	3.8	0.5	1.9
Polyethylene terephthalate	500		0.040	0.035	2.9	2.6	7.5
Polyurethane foam, TDI, not fire ret.	510	440	0.030	0.020	6.6	1.5	10.
Polyurethane foam, TDI, fire ret.	550	390	0.025	0.015	9.8	1.7	10
Allyl alcohol derivative							
sample 1, CR-39	510		0.035	0.020	5.6	3.6	<10
sample 2, CR-39	500		0.035	0.060	1.9	6.7	>10
sample 3, CR-149, glass (65/35)	540		0.345	1.600	0.1	0.2	< 0.1
Phenol furfural	530		0.025	0.010	15.2	3.9	>10
Phenol furfural, 1.5% glycerol monooleate and 1% K_2CO_3	520	310	0.025	0.010	15.5	4.0	>10
Alkyd molding compound, mineral filler							
sample 1, not self extinguishing	500	270	0.155	0.120	0.2	<0.1	< 0.1
sample 2, self extinguishing	510	270	>2.00	>8.32	0.1		< 0.1

(Concluded)

SECTION 4.4. TESTS FOR IGNITABILITY

Ignitability may be defined as the facility with which a material or its pyrolysis products can be ignited under given conditions of temperature, pressure, and oxygen concentration. This characteristic provides a measure of fire hazard in that a material which has an ignition temperrature significantly higher than that of another material is less likely to contribute to a conflagration, all other factors being the same in both cases. For example, if a fire wall could be relied upon to keep the temperature on the side away from the fire below 300°C. for a stated period of time, a material that ignited at 200°C. would be a potential hazard and a material that ignited at 400°C. would not be a potential hazard. Some measures of ease of ignition are autoignition temperature, flash ignition temperature, and ignition sensitivity.

Almost any material can be made to ignite, given enough heat, enough oxygen, and enough time. Ease of ignition can therefore be measured by the amount of heat required under fixed conditions of oxygen and time, by the amount of oxygen required under fixed conditions of heat and time, or by the amount of time required under fixed conditions of heat and oxygen. The most simple tests provide fixed conditions of heat, oxygen, and time, and the specimen either ignites or does not ignite under those fixed conditions.

There are four possible sources of heat energy which can result in ignition: chemical heat energy, such as heat of combustion; electrical heat energy, such as heat from arcing; mechanical heat energy, such as heat from friction; and nuclear heat energy, such as heat of fission. In general, only two sources of energy are used in tests for ease of ignition: chemical heat energy, in the form of a direct flame or heated object, and electrical heat energy, in the form of resistance elements or arcs.

Many tests provide a measure of both ease of ignition and surface flammability. Some tests have, in essence, become measures of surface flammability for easily ignited materials, and measures of ease of ignition for materials that are difficult to ignite. For convenience in discussion, all tests which provide some measure of surface flammability will be described in Section 4.5; in general, this group includes all tests which provide a measurable distance available for relatively free flame travel.

Both the ASTM E 136 (4.301) and ASTM D 1929 (4.302) tets employ a vertical furnace tube 254 mm (10 in) long with 102 mm (4 in) inside diameter, heated by electrical current passing through nichrome wire in an asbestos sleeve wound around the tube, and an inner refractory tube 254 mm (10 in) long with 76 mm (3 in) inside diameter, inside which the specimen is placed. Air is admitted at a controlled rate, and its temperature measured by means of thermocouples.

In the ASTM E 136 test (4.301), a specimen 51 mm (2 in) long, 38 mm

(1.5 in) wide, and 38 mm (1.5 in) thick is placed in a stream of air at 750°C. (1382°F.) moving at 3 m (10 ft) per minute. A material is considered noncombustible by this test if specimen temperatures do not increase more than 30°C. (54°F.) and there is no flaming after the first 30 seconds.

In the ASTM D 1929 test (4.302), a 3 gram specimen is exposed to air at successively higher temperatures until ignition is observed. Flash-ignition temperature is defined as the lowest initial temperature of air passing around the specimen at which a sufficient amount of combustible gas is evolved to be ignited by a small external pilot flame. Self-ignition temperature is defined as the lowest initial temperature of air passing around the specimen at which, in the absence of an ignition source, the self-heating properties of the specimen lead to ignition or ignition occurs of itself, as indicated by an explosion, flame, or sustained glow. This test is also known as the Setchkin ignition test. Data for various materials are given in Table 2.5.

The classic approach to the study of ignition behavior of materials involves the measurement of time to ignition at different heat flux levels, and the work at the University of Oklahoma has been excellent and extensive (4.303). One benefit of high-flux studies using the NBS chamber has been the availability of compact heaters capable of producing ignition by the application of unidirectional heat flux (4.304, 4.305). The Mellen high-flux heater has been used for ignitability studies at th University of San Francisco (4.306). These electrical heaters provide advantages in compactness, convenience, and operating safety over gas-fueled open flames; the user, however, should be aware of the differences which could result from the selection of any energy source other than flame radiation.

The USF ignitability test employs a 50/50 platinum/rhodium wire heater to provide heat flux up to 13 W/cm² (4.307). A specimen 76.2 mm (3 in) square is supported vertically in a frame such that an area of front surface 65.1 mm (2-9/16 in) square is exposed to a specified level of heat flux. Time to ignition is recorded. Some data obtained using this test (4.307) are presented in Table 3.2.

It is readily apparent that time to ignition generally decreases with increasing heat-flux. Since heat flux is a function of distance from the heat source and is reduced by intervening materials, placement and design of the material in the system can often determine whether or not ignition will occur in a particular fire situation.

The ASTM D 229 test for electrical insulation materials (4.308) consists of two parts. In part I, a horizontal specimen measuring 127 by 13 mm (5 by ½ in) is ignited at one end by a burner flame applied at a 30° angle for 30 seconds. In part II, a vertical specimen measuring 254 by 13 by 13 mm (10 by ½ by ½ in) is exposed to the heat from a coil of nichrome resistance wire maintained at 860 ± 5°C. by the passage of approximately 55 amp. of electrical current, with continuous sparking provided to ignited evolved gases.

Method 2023 of Federal Test Method Standard No. 406 (4.309) employs a vertical specimen, measuring 5 by ½ by ½ in., surrounded by heater coils supplied with 55 map. of electrical current.

An Underwriters Laboratories ignition test (4.310) employs a high-temperature glass flask surrounded by a molten-alloy bath heated by an electric furnace. The flask is conical, with a flat bottom 6.0 cm in diameter, 11.4 cm high, and 2.8 cm in diameter at the top, giving a total volume of 160 ml and a surface area to volume ratio of 1.1 cm^{-1}. Specimens are dropped into the flask after temperature has reached equilibrium at a selected value, and are checked for ignition. Because this test provides conditions of natural convection inside the flask rather than controlled air flow, its results are not necessarily the same as those obtained with the Setchkin test.

Four Underwriters Laboratories tests (4.311) are used to evaluate ease of ignition of materials intended for electrical applications. The hot-wire ignition test employs specimens measuirng 12.7 by 1.27 cm (5 by 0.5 in) in three thicknesses, 1.6, 3.2, and 6.4 mm (1/16, 1/8, and ¼ in), wrapped with five turns of No. 24 wire spaced 6.4 mm (¼ in) between turns. The wire is brought to red heat by applying a 65 watt current for a maximum of 5 minutes. The high-current-arc ignition test subjects a specimen to 40 applications per minute, for a maximum of 5 minutes, of an electric arc fed by a 32.7 amp. current from a 240 volt power supply. The high-voltage-arc ignition test subjects a specimen to a 5200 volt arc for a maximum of 2 minutes. The high-voltage-arc tracking test subjects a specimen to a 5200 volt arc which is repeatedly extinguished and reestablished by changing the arc distance, until the conductive path is 50.8 mm (2 in) in length or until 2 minutes of arching time has elapsed.

Method 4011 of Federal Test Method Standard No. 406 (4.312) measures the resistance of a sepcimen to arcs generating up to 56 watts of heat.

SECTION 4.5. TESTS FOR FLAME SPREAD

Flame spread may be defined as the rate of travel of a flame front under given conditions of burning. This characteristic provides a measure of fire hazard, in that flame spread can transmit fire to more flammable materials in the vicinity and thus enlarge a conflagration, even though the transmitting material itself contributes little fuel to the fire. Some measures of flame spread are burning rate or combustion rate, burning extent or distance of flame travel, flame spread factor, and flame height.

Tests for flame spread can be classified and numerically described in Cartesian coordinates by the angle formed by the exposed surface with the horizontal, surface angle θ is 0° for a horizontal material burning on its upper surface with the flame front moving away from the origin, as in ASTM D 1692, rotating this surface through the first quadrant until θ becomes 90° gives a burning vertical surface on which the flame front travels upward, as in ASTM D 3014. Further rotation through the second quadrant until θ is 180° gives a horizontal material burning on its lower surface, as in ASTM E 84. Further rotation through the third quadrant until θ is 240° gives a burning surface such that the flame front travels downward at an angle of 30° from the vertical, as in ASTM E 162.

The surface angle θ determines the extent to which hot combustion gases will preheat the surface ahead of the advancing flame front. Because these gases travel directly upward unless obstructed, the degree to which preheating by this mechanism occurs is a function of sin θ, which is the preheat factor. Where forced air flow is controlling, rather than natural air flow, then the preheat factor becomes a function of the direction of forced air flow rather than surface angle θ, forced air flow in the direction of flame front travel giving a preheat factor of essentially 1.0. A preheat factor of 1.0 represents the most severe conditions from the viewpoint of preheat, and a preheat factor of −1.0 the least severe conditions.

The surface angle θ also determines the extent to which hot combustion gases will promote penetration into the material by exerting pressure perpendicular to the exposed surface, and the extent to which heavy gases will move perpendicular to the exposed surface. The degree to which these phenomena occur is a function of −cos θ, the buoyancy factor. Where the combustion gases are restrained by confinement in a limited space, the buoyancy factor must be adjusted to a more realistic value.

The extent to which the heat of the exposure fire and the heat of combustion are concentrated in proximity to the exposed surface is a function of the dimensions of the enclosure relative to the dimensions of the exposed surface. The effective enclosure volume per unit of surface area is the concentration factor, a measure of the severity of the test. Severity of exposure, however, does not increase indefinitely with increasing concentration factor.

The ASTM E 84 tunnel test (4.501) requires a specimen 7.62 by 0.496 m (300 by 19.5 in), mounted face down so as to form the roof of a tunnel 7.62 by 0.445 by 0.305 m (300 by 17.5 by 12 in). The fire source, two gas burners 305 mm (12 in) from the fire end of the sample and 190 mm (7.5 in) below the surface of the sample, is adjusted so that a test sample of select-grade red oak flooring would spread flame 5.94 m (19.5 ft) from the end of the igniting fire in 5.5 min ± 15 sec. The end of the igniting fire is considered as being 1.37 m (4.5 ft) from the burners, this flame length being due to an average air velocity of 73.2 ± 1.5 m/min (240 ± 5 ft/min). Materials are rated on a flame spread classification with red oak as 100. Some test data are presented in Table 3.4.

The ASTM E 162 radiant panel test (4.502) employs a radiant heat source consisting of a 305 by 457 mm (12 by 18 in) vertically mounted porous refractory panel maintained at 670 ± 4°C (1238 ± 7°F). A specimen measuring 152 by 457 mm (6 by 18 in) is supported in front of it with the 457 mm (18 in) dimension inclined 30 degrees from the vertical. A pilot burner ignites the top of the specimen, 121 mm (4¾ in) away from the radiant panel, so that the flame progresses downward along the underside exposed to the radiant panel. A flame spread factor is calculated from the times required for the flame front to reach successive marks 3 in apart along the specimen surface. Some test data are presented in Table 4.5.

The ASTM E 648 or National Bureau of Standards flooring radiant panel test (4.503) employs a gas-fueled porous refractory radiant panel, with a radiation surface of 305 by 457 mm (12 by 18 in), inclined at 30 degrees to and directed downward at a horizontally mounted specimen 250 by 1005 mm (9.9 by 41.4 in), with an enclosure 1400 by 500 by 710 mm (55 by 19.5 by 28 in) above the test specimen. The radiant panel is maintained at a temperature of 500°C (932°F), and the radiant energy flux is plotted as a function of distance along the specimen plane to produce a

Table 4.5. Flame Spread and Smoke Characteristics of Materials as Measured by the ASTM E 162 Test

Material	Thickness mils	Flame Spread Index
Red oak	750	99
Hardboard	218	136
Fiberboard, unfinished	500	236
Polystyrene, extruded	66	355
Polystyrene, tile	68	224
Polymethyl methacrylate, FR	125	376
Polyvinyl chloride, rigid	147	9.6
Polyvinyl chloride, rigid, FR	147	3.2
Polyvinyl chloride, flexible, on cotton	21	89
Polyvinyl chloride, flexible, FR, on cotton	18	4.5
Phenolic, laminate	63	107
Polyester, 21% glass reinforced	62	239
Polyester, 27% glass reinforced	85	154
Polyester, 27% glass reinforced, FR	95	66

flux profile curve. The critical radiant flux at flame-out or the point of farthest advance of the flame front is calculated from the flux profile curve. Some data obtained using this test are presented in Table 4.6.

The ASTM E 286 8-foot tunnel test (4.504) employs a specimen 8 feet long and 14 inches wide, mounted horizontally so as to form the roof of a tunnel so as to give a lengthwise slope of 6 degrees and a sidewise tilt of 30 degrees.

The Union Carbide Corporation four-foot tunnel (4.505) employs a specimen measuring 47½ inches long and 7½ inches wide, mounted horizontally so as to form the roof of a tunnel 6¾ inches deep. The heat supply rate is 325 Btu per minute, compared to initial trial rates of 3400 Btu per minute for the 8-foot tunnel, and 5000 Btu per minute for the 25-foot tunnel.

The Monsanto Company two-foot tunnel test (4.506) employs a specimen measuring 24 by 4 inches, inclined 28 degrees from the horizontal so as to form the roof of the enclosed tunnel, and ignited by a Fisher 3-900 burner supplied with a constant 3 oz per sq. in. natural gas flow. The test method provides flame spread rating, fuel contribution, afterflaming, afterglowing, intumescence, insulative value, and smoke contribution.

Table 4.6. Flame Spread Characteristics of Materials as Measured by the ASTM E 648 Test

Material	**Weight oz/yd²**	**Critical Radiant Flux, W/cm²**
Residential carpets (no preheat)		
nylon, level loop		0.76 ± 0.13
wool, plush		0.64 ± 0.10
nylon 6, cut pile		0.59 ± 0.09
acrylic, level loop		0.47 ± 0.06
polyester		0.40 ± 0.10
acrylic		0.23 ± 0.05
Contract carpets (2-min prepheat)		
nylon 6/6, level loop, tufted, jute backing	28	did not ignite
wool, level loop, velvet, latex backing	46	0.83 ± 0.09
nylon 6/6, cut loop, woven, latex backing	34.7	0.67 ± 0.13
polyester, level loop, tufted, jute backing	42	0.66 ± 0.08
acrylic/nylon, level loop, tufted, foam backing	37	0.64 ± 0.12
acrylic, level loop, tufted, jute backing	42	0.60 ± 0.08
nylon 6, level loop, tufted, foam backing	16	0.19 ± 0.04
polypropylene, level loop, tufted, jute backing*	28	0.10

*With carpet cushion pad

The Schlyter test (4.507) employs two specimens, each 12 in. wide and 31 in. high, held in a vertical position with their faces 2 inches apart. The bottom of the assembly is subjected for 3 minutes to the flame from either a bunsen burner with wing top delivering 37 Btu per minute, or a No. 4 meker burner with special T-head delivering 291 Btu per minute.

The Pittsburgh-Corning 30-30 tunnel (4.508) employs a specimen measuring 30 by 3-7/8 inches inclined 30 degrees from the horizontal so as to form the roof of the enclosed tunnel. The flame source, is a Fisher 3-900 burner supplied with 2600 cc per minute of natural gas, giving a 6-inch flame providing about 90 Btu per minute.

The ASTM D 635 test (4.509) is perhaps the most widely used small-scale test for plastic materials. The specimen, measuring 125 by 12.5 mm (5 by 0.5 in) by the supplied thickness, is held with the 125 mm (5 in) dimension horizontal and the 12.5 mm (0.5 in) dimension inclined at a 45 degree angle. One end is contacted for 30-second periods with a 25 mm (1 in) high blue flame from a 3/8 in. diameter barrel bunsen burner.

The ASTM D 1692 test (4.510) has been discontinued but is still referenced in some older military specifications. The specimen, measuring 150 by 50 mm (6 by 2 in) by up to 13 mm (0.5 in) thick, is supported on a horizontal hardware-cloth support with the 13 mm (0.5 in) dimension vertical. One end is contacted for 60 seconds with a 38 mm (1.5 in) high blue flame with a 6.5 mm (0.25 in) inner cone from a 9.5 mm (3/8 in) diameter barrel bunsen burner fitted with a 48 mm (1-7/8 in) wide wing top. The test chamber measures 300 by 600 m by 760 mm high, with 25 mm high ventilating openings around the bottom.

The UL 94 standard (4.511) employs essentially the ASTM D 635 test as its horizontal burning test for 94HB classification, and the ASTM D 1692 test as its horizontal burning tesst for 94HBF, 94HEF-1, or 94HEF-2 classification.

The ASTM D 3014 test (4.512) contacts the bottom of a vertical specimen, measuring 254 by 19 by 19 mm (10 by 0.75 by 0.75 in), for 10 seconds with a 960°C. (1760°F.) flame from a bunsen burner, the specimen being enclosed in a vertical steel chimney lined with reflective aluminum foil to increase the severity of exposure.

The ASTM D 568 test (4.513) employs a test shield measuring 300 by 300 mm by 760 mm high, open at the top. A specimen measuring 450 by 25 mm (18 by 1 in) by the supplied thickness is held vertically, and the bottom contacted for 15 seconds with a 25 mm (1 in) high flame from a bunsen burner.

The ASTM D 1433 test (4.514) employs a specimen measuring 228 by 76 mm (9 by 3 in), mounted at a 45 degree angle in a test chamber. The bottom end is exposed to a 13 mm (0.5 in) flame from a No. 22 hypodermic needle jet.

The ASTM D 757 test (4.515) employs a specimen measuring 121 by 3.17 by 12.7 mm (5 by 1/8 by ½ in). One end is placed in contact for 3 minutes with a Globar element maintained at 950°C. (1742°F.).

Federal Test Method Standard No. 406 (4.516) describes two test methods similar to ASTM tests. Method 2021 is similar to ASTM D 635. Method 2022 is similar to ASTM D 568, except that ignition by a burner flame is replaced by ignition with a pyroxylin fuse or benzene drop.

The ASTM D 1360 test for paints (4.517) employs a cabinet measuring 13¼ by 9 by 18¼ inches, and a specimen measuring 6.4 by 152 by 305 mm (¼ by 6 by 12 in). The ignition source is 5 ml of absolute ethyl alcohol in a cup under the specimen.

The ASTM D 1361 test for paints (4.518) employs a fire shield measuring 203 by 279 by 762 mm (8 by 11 by 30 in), open at the top, and a vertical stick measuring 25.4 by 25.4 by 406 mm (1 by 1 by 16 in) painted on all four sides. The stick is wrapped with a wick soaked with 4 ml of ethyl alcohol and ignited. This test has been discontinued.

The ASTM D 1230 test for fabrics (4.519) employs a chamber measuring 268 by 216 mm (14.5 by 8.5 in) by 256 mm (14 in) high, and a specimen measuring 51 by 152 mm (2 by 6 in) mounted at a 45 degree angle. The ignition source is a 16 mm (5/8 in) flame from a 26 gage hypodermic needle, applied for 1 second.

The ASTM D 2859 test for floor coverings (4.520) employs a test chamber measuring 305 by 305 by 305 mm (12 by 12 by 12 in) and a specimen measuring 230 by 230 mm (9 by 9 in) placed on the floor of the chamber. The ignition source is an Eli Lilly No. 1588 methenamine tablet.

The ASTM D 2859 test is essentially the same as that described in flammability standards DOC FF 1-70 (4.521) and DOC FF 2-70 (4.522), for carpets and rugs and for small carpets and rugs, respectively.

Federal Test Method Standard No. 501a for nontextile resilient floor coverings (4.523) describes two test methods. In Method 6411, a specimen measuring 31.5 by 7 inches is mounted horizontally and exposed to flames from 4 burners supplied with 9.6 cu. ft./hr. propane and 150 cu.ft./hr. air for 240 seconds, in a hood consisting of a horizontal flue measuring 38 by 6 inches and a vertical flue measuring 18 by 8 by 6 inches. In Method 6421, a specimen measuring 18 by 6 inches, inclined at a 60 degree angle, is exposed to a 670°C. radiant panel and ignited at the upper end.

The Underwriters Laboratories floor furnace (4.524) consists of a test chamber measuring 10 by 22 inches by 10½ feet. A specimen measuring 24 by 96 inches is mounted horizontally on the floor, and one end is exposed to a gas burner flame delivering 500 Btu/min. and an air velicity of 100 ft./min. for 12 minutes.

The eight-foot corner test developed at University of California (4.525) employs enough specimen material to construct a corner consisting of two walls each 6 feet wide and 8 feet tall, topped with a 6-foot-square ceiling. Various ignition sources have been used, such as a 2800-gram quantity of milk cartons.

The Factory Mutual corner test (4.526) employs enough sample material to construct a corner consisting of two walls both 24 ft. 9 in. high, one wall 50 ft. in length and the wall 37 ft. 9 in. in length. The igniting fire is normally produced by burning a 5 ft. high pile of 4 by 4 ft. wood pallets weighing about 750 pounds with an average moisture content of less than 8 percent.

The ASTM E 108 test for roof coverings (4.527) describes three test procedures. In the intermittent flame test, a specimen measuring 3 ft. 4 in. by 4 ft. 4 in., mounted at a specified slope, is exposed to flame at 760°C. (1400°F.) for Classes A and B, or 740°C. (1300°F.) for Class C, and a 12 mph air current, intermittently according to a specified sequence. In the flame spread test, a specimen measuring 3 ft. 4 in. by 13 ft., mounted at a specified slope, is exposed to flame at 760°C. (1400°F.) for 10 minutes for Classes A and B, or 704°C. (1300°F.) for 4 minutes for Class C, and a 12 mph air current. In the burning brand test, a specimen measuring 3 ft. 4 in. by 4 ft. 4 in., mounted at a specified slope, is exposed to a 12 mph air current and a burning brand, which weighs 2000 g. for Class A, 500 g. for Class B, or 9.25 g. for Class C. Class A and Class B brands are made of Douglas fir, Class C brands are made of white pine.

The ASTM E-69 fire tube test for treated wood (4.528) employs a specimen mesuring 9.5 by 19 by 1016 mm (3/8 by ¾ by 40 in), mounted vertically in a 3 inch diameter fire tube and exposed to a 279 mm (11 in) high blue flame from a burner.

The ASTM E 160 crib test for treated wood (4.529) employs a crib test specimen consisting of 24 pieces each measuring 13 by 13 by 76 mm (½ by ½ by 3 in), exposed to a 254 mm (10 in) high blue flame from a meker burner.

The Underwriters Laboratories standard UL 214 (4.530) describes two tests. In the small-flame test, a specimen measuring 2¾ by 10 in. is mounted vertically and the lower end is exposed to a 1.5 in. long vertical flame from a burner for 12 seconds, in a test chamber measuring 12 by 12 by 30 in. In the large-flame test, specimens in single sheets (measuring 5 by 84 in) or folds (measuring 25 mm by 84 in, folded in 5-inch pleats) are mounted vertically, and the lower end exposed to an 11-inch long flame from a burner for 2 minutes, in a test chamber measuring 12 by 12 by 84 in.

Federal Test Method Standard No. 191 (4.531) describes six test methods for textiles. These are summarized as follows:

Method 5900: specimen 7 by 10 in., horizontal, exposed to vertical flame from 0.3 ml. anhydrous ethyl alcohol

Method 5903: specimen 2¾ by 12 in., vertical, lower end exposed to 1½ in. long vertical flame from burner for 12 seconds, test chamber measuring 12–14 by 12–14 by 30 in.

Method 5904: specimen 2 by 5 in., vertical, bottom edge exposed to vertical flame from ¾ in. diameter paraffin candle.

Method 5906: specimen 4½ by 12½ in., horizontal, one end exposed to 1½ in. long vertical flame from burner, test chamber measuring 15 by 8 by 14 in.

Method 5908: specimen 2 by 6 in., inclined at 45 degree angle, lower end exposed to 5/8 in. flame from hypodermic needle for 1 second.

Method 5910: specimen 1 by 6 in., inclined at 30 degree angle, lower end exposed to flame from safety-book match for 5 seconds.

The fabric flammability test of greatest commercial importance is the test described by the DOC FF 3-71 flammability standard for children's sleepwear (4.532). It employs a vertical specimen measuring 3.5 in. by 10 in., mounted in a test chamber measuring 12 by 12 by 31 in., and exposed to a 1.5 in. long vertical flame from a burner for 3.0 ± 0.2 seconds.

The Motor Vehicle Safety Standard No. 302 test (4.533) employs a specimen measuring 4 by 14 in., mounted horizontally in a test chamber measuring 15 by 8 by 14 in. One end is exposed to a 1.5 in. high vertical flame from a burner. This test is essentially the same as SAE J 369, General Motors Test Method 32-12, and Chrysler LP-463KC-13-01.

The ASTM D 3675 radiant panel test for flexible cellular plastics (4.534) is a modification of the ASTM E 162 test to make it suitable for flexible foams. Modifications include a wire mesh screen to support the sample.

The Upjohn small corner test (4.535) employs a corner configuration 4 feet on each side and 2 feet high, and is designed to predict results in the Factory Mutual full scale corner test.

SECTION 4.6. TESTS FOR HEAT RELEASE

Heat release may be defined as the heat produced by the combustion of a given weight or volume of material. This characteristic provides a measure of fire hazard in that a material which burns with the evolution of little heat per unit quantity burned will contribute appreciably less to a conflagration than a material which generates large amounts of heat per unit quantity burned. Some measures of heat release are heat evolution factor and fuel contribution index.

Tests for heat release provide a measure of the amount of heat produced by the material in the process of burning. Such tests generally provide information, not on the absolute amount of heat, but on the relative amount of heat produced, or on the equivalent amount of fuel required to produce similar results, for purposes of comparison with other materials.

The Factory Mutual calorimeter test (4.601) employs a specimen measuring 4½ ft. by 5 ft., of which an area 4 ft. by 4 ft. is exposed in the roof of a furnace 17½ ft. long, 4 ft. wide, and 3¾ ft. high. Gasoline is used to fuel the main fire exposure burners while propane is used as the evaluating fuel. After the specimen is burned using only the main fire exposure burners, it is replaced with a noncombustible cover and the evaluation burners are adjusted to reproduce the flue temperature-time curve produced by burning the specimen. The heat added through the evaluation burners is considered equal to the heat released by the specimen.

The Ohio State University release rate apparatus (4.602) employs a chamber 890 by 410 by 200 mm (35 by 16 by 8 in), with a pyramidal top section 395 mm (15.5 in) high connecting to the outlet. A radiant heat source, consisting of four silicon carbide Globar Type LL elements, is used to generate heat flux up to 13 W/cm². Specimens 150 by 150 mm (6 by 6 in) are tested in vertical orientation, and specimens 110 by 150 mm (4.5 by 6 in) are tested in horizontal orientation. A radiation reflector is used for horizontally mounted specimens. The total air flow of 0.04 m³/3 (84 ft³/min) leaves the apparatus through a rectangular exhaust stack 133 by 70 mm (5.25 by 2.75 in) in cross section and 254 mm (10 in) high. The temperature difference between the air entering and the air leaving the apparatus is measured by a thermopile having 3 hot junctions spaced across the top of the exhaust stack and 3 cold junctions located in the pan at the bottom. Some OSU heat-release data are presented in Table 3.5.

The crucible test method developed by CSTB in France (4.603) involves a controlled rate pyrolysis of the test material at 800 to 900°C. in a crucible. When the concentration of the combustible gases evolved reaches the lower limit of flammability at the outlet, they are ignited with a small pilot flame, and the heat produced is measured by means of a calorimeter filled with water. The furnace is built with a 104/108 mm diameter steel tube, 93 mm high, in which refractory cement is cast to provide a truncated conical void space 40 mm in diameter at the bottom, 80 mm on the top, with a ledge 29 mm from the bottom for the crucible support. The crucible is made of 18/10 stainless steel, 1 mm thick, with a capacity of 85 cu cm, height of 50 mm, and external diameter of 53 mm. The calorimeter is made of copper, 1 mm thick, 100 mm high, 83 mm external diameter, with a bottom cavity 48 mm in diameter and 50 mm high.

The ASTM E 84 25-foot tunnel test (4.501) and the ASTM E 162 radiant panel test (4.502) are essentially tests for surface flame spread, but they provide measures of heat release based on the temperature rise in the gases produced.

SECTION 4.7. TESTS FOR FIRE ENDURANCE

Fire endurance may be defined as the resistance offered by the material to the passage of fire, normal to the exposed surface over which

Table 4.7. Smoke-Producing Characteristics of Materials as Measured by the Arapahoe Smoke Test

Material	Percent Smoke Based on Initial Weight	Percent Smoke Based on Weight Loss
Hardwoods		
Beech	0.05	0.08
Yellow birch	0.09	0.14
Red oak	0.11	0.18
Aspen poplar	0.13	0.20
Softwoods		
Southern yellow pine	0.08	0.12
Douglas fir	0.14	0.24
Eastern white pine	0.17	0.27
Western hemlock	0.20	0.33
Western red cedar	0.23	0.35
Cellulosic boards		
Hardboard	0.06	0.08
Medium density hardboard	0.11	0.16
Cellulose fiberboard, core board	0.57	0.75
Plastics		
Polymethyl methacrylate	0.08	0.19
Polyvinyl chloride flooring	0.21	6.23
Acrylic, unidentified	0.33	0.47
Linoleum	0.52	1.53
Polychloroprene rubber, filled, FR	0.80	8.72
Polycarbonate, 2	0.89	13.34
Polyvinyl chloride, rigid	1.33	10.52
Polycarbonate, 1	1.34	11.96
Polyvinyl chloride/acrylic	1.38	7.60
Polypropylene, FR	1.64	13.42
Polyester, brominated, fiberglass reinforced	1.70	14.52
Polyvinyl chloride, flexible, FR	2.36	12.74
Acrylonitrile-butadiene-styrene, FR	4.02	20.54
Polystyrene	4.86	12.71
Styrene-butadiene rubber, filled, FR	6.63	10.51

flame spread is measured. This characteristic provides a measure of fire hazard, in that a material which will contain a fire represents more protection than one which will give way before the same fire, all other factors being the same in both cases. Some measures of fire endurance are penetration time and resistance rating.

More tests for fire endurance are more concerned with complete systems than with individual materials, because it is widely recognized that the performance of the individual materials comprising a system is not necessarily indicative of the performance of the system as a whole.

The ASTM E 119 test standard (4.701) for building construction and materials, provides for exposure of various structural components to a standard fire, the character of which is determined by a standard time-temperature curve. The main points on the curve are:

538°C. (1000°F.) at 5 minutes
704°C. (1300°F.) at 10 minutes
843°C. (1550°F.) at 30 minutes
927°C. (1700°F.) at 1 hour
1010°C. (1850°F.) at 2 hours
1093°C. (2000°F.) at 4 hours
1260°C. (2300°F.) at 8 hours or over

The performance is defined as the period of resistance to standard exposure elapsing before the first critical point in behavior is observed, and is expressed in the time ratings, such as 2-hour and 4-hour.

For bearing walls and partitions, the area exposed to fire shall not be less than 100 sq. ft. (9 sq. m), with neither dimension less than 9 ft. (2.7 m). For a period equal to that for which classification is desired, the wall or partition should not permit passage of flame or gases hot enough to ignite cotton waste, should sustain specified loads during and after test, and should not permit the temperature rise on the unexposed surface to exceed 139°C. (250°F.).

For nonbearing walls and partitions, the area exposed to fire shall not be less than 100 sq. ft. (9 sq. m), with neither dimension less than 9 ft. (2.7 m). For a period equal to that for which classification is desired, the wall or partition should not permit passage of flame or gases hot enough to ignite cotton waste, and should not permit the temperature rise of the unexposed surface to exceed 139°C. (250°F.).

For columns, the length of the column shall not be less than 9 ft. (2.7 m). For a period equal to that for which classification is desired, the column should sustain a specified load.

For protected structural steel columns, the length of the column shall not be less than 8 ft. (2.4 m). For a period equal to that for which classification is desired, the average temperature of the steel should not exceed 538°C. (1000°F.) at any of four levels, and the temperature at any of at least three points on any of the four levels should not exceed 649°C. (1200°F.).

For floors and roofs, the area exposed to fire shall not be less than 180 sq. ft. (16 sq. m), with neither dimension less than 12 ft. (3.7 m). For a period equal to that for which classification is desired, the construction should sustain a specified load, should not permit passage of flame or gases hot enough to ignite cotton waste, and should not permit the temperature rise of the unexposed surface to exceed 139°C. (250°F.). For specimens employing steel structural members, the temperature of the steel shall not have exceeded 704°C. (1300°F.) at any location during the classification period, nor shall the average temperature recorded by four thermocouples at any section have exceeded 593°C. (1100°F.). For specimens employing concrete structural members, the average temperature of the tension steel at any specimen shall not have exceeded 427°C. (800°F.) for cold drawn prestressing steel or 593°C. (1100°F.) for reinforcing steel during the classification period.

For loaded beams, the length of beam exposed to fire shall not be less than 12 ft. (3.7 m). A section of a representative floor or roof construction not more than 7 ft. (2.1 m) wide may be included with the test specimen. For a period equal to that for which classification is desired, the specimen shall have sustained the specified load.

For protected structural steel beams and girders, the length of beam or girder exposed to fire shall not be less than 12 ft. (3.7 m) and the exposed section of floor construction shall not be less than 5 ft. (1.5 m) wide. For a period equal to that for which classification is desired, the average temperature of the steel at any one of four sections should not exceed 538°C. (1000°F.), and the temperature at any one of at least four points at any of the four sections should not exceed 649°C. (1200°F.).

For ceiling constructions, the area exposed to fire shall not be less than 180 sq. ft. (16 sq. m), with neither dimension less than 12 ft. (3.7 m). For a period equal to that for which classification is desired, the ceiling should not permit the passage of flame, should not permit the temperature rise of combustible supports to exceed 139°C. (250°F.) at points of contact or at adjacent surfaces, and should not permit the average temperature of noncombustible supports to exceed 538°C. (1000°F.).

For protected combustible framing or facings, the area exposed to fire shall not be less than 100 sq. ft. (9 sq. m) for wall and partition protection and not less than 180 sq. ft. (16 sq. m) for floor protection, neither dimension being less than 9 ft. (2.7 m) and 12 ft. (3.7 m), respectively. For a period equal to that for which classification is desired, the protection should not permit ignition of the materials protected, and should not permit the temperature rise at points of contact with the protected structural members to exceed 139°C. (250°F). The permissible temperature rise is 181°C. (325°F.) for members closely imbedded on three sides in non-combustible materials.

The standard time-temperature curve described in ASTM E 119 is also used to define the exposure fire for the ASTM E 152 test for door assemblies (4.702) and the ASTM E 163 test for window assemblies (4.703).

Similar severity of exposure is involved in the Union Carbide Corporation tests for thermal insulation materials on pipes (4.704) and vessels (4.705). In these tests the flow of gasoline fuel is controlled so that the fire exposure approximates that of the ASTM E 119 standard time-temperature curve.

The Factory Mutual heat damage test (4.706) employs a horizontal specimen measuring 16 by 16 in. of the roof or wall construction to be tested. The specimen forms the roof of a small furnace, which is heated so that the temperature one inch below the bottom face of the specimen increases as follows:

for roof insulations:	550°F. at 10 min.
	750°F. at 20 min.
	850°F. at 30 min.
for wall insulations:	425°F. at 5 min.
	475°F. at 10 min.
	500°F. at 15 min.
	500°F. at 20 min.

The UL 181 test for air ducts (4.707) employs a horizontal specimen 19 in. sq., subjected to an 8-pound static load with a bearing surface on the sample of 1 by 4 in. The specimen forms the roof of the test furnace. Gas burners 20 in. below the specimen are fired such that a thermocouple 3 in. below the underside of the specimen indicates 1400°F. for a 30-minute period.

The SS-A-118b test (4.708) employs a specimen measuring 36 by 36 in., supported in a horizontal position with a 30 by 30 in. area exposed,

28¾ in. above a burner delivering either 28,000 Btu during a 20-minute test of 60,000 Btu during a 40-minute test.

The Bureau of Mines flame penetration test (4.709) employs a horizontally mounted specimen measuring 6 by 6 in. by 1 in., and measures the time required for the 1-inch thickness to be penetrated by a 2150°F. propane-air flame from a vertical Bernz-O-Matic pencil-flame burner. An earlier version (4.710) employed a vertically mounted specimen and a horizontal flame at 1930°F.

The NASA Ames T-3 thermal test (4.711) employs specimens measuring 12 by 12 by 2 in., mounted in one of three test areas in the chamber. The flames from an oil burner supplied with approximately 1½ gallons per hour of JP-4 jet aviation fuel provide heat flux on these three areas of 5.5 to 11, 9 to 16, and 22 Btu per sq. ft. per second.

The ICI test for pipe insulation (4.712) employs a horizontal pipe 107 to 120 cm long, filled with an inert liquid and covered with the insulation material to be tested. The fire source is two liters of kerosene poured into a metal tray below the test pipe and ignited.

SECTION 4.8. TESTS FOR EASE OF EXTINGUISHMENT

Ease of extinguishment may be defined as the facility with which burning can be extinguished in the case of a specific material. This characteristic provides a measure of fire hazard, in that a material which requires more effort to extinguish is more likely to prolong the fire than one which is easily extinguished, all other factors being the same in both cases.

The ASTM D 2863 oxygen index test (4.801) employs a vertical heat-resistant glass tube at least 75 mm in diameter and at least 450 mm high, in which a specimen 70 to 150 mm long, 6.5 mm wide, and 3.0 mm thick is held vertically by a clamp at its bottom end. A mixture of oxygen and nitrogen of known composition is metered into the bottom of the tube, passing through a bed of glass beads 3 to 5 mm in diameter and 80 to 100 mm deep at the bottom to smooth the flow of the gas. The gas flow rate in the column is 4 ± 1 cm/sec. The specimen is ignited at its upper end with an igniting flame which is then withdrawn, and the atmosphere which just permits steady burning down of the specimen is determined. The oxygen index is the minimum concentration of oxygen in an oxygen-nitrogen mixture which will just permit the sample to burn, that is, burn 3 minutes or 50 mm. Data for various materials are given in Table 2.6.

Table 4.8. Smoke-Producing Characteristics of Materials as Measured by the ASTM D 2843 Test

Material	Thickness mils	Maximum Light Absorption pct
Polyethylene, UCC DXM-100	250	9
Polyethylene, UCC DFDA-6311	125	82
Polyethylene, UCC DHDA-1811	125	79
Polyethylene, UCC DMDA-7075	125	49
Polypropylene, UCC JMD-8500	250	20
Polyvinyl chloride, UCC QCA-2460	15	39
	20	40
	40	87
Polyvinyl chloride, UCC QYTQ	250	100
Polyvinyl butyral, UCC XYHL	250	11
Polystyrene, UCC SMD-3500	250	98
Styrene-acrylonitrile, UCC C-11	250	100
ABS, Cycolac	250	100
Polysulfone, UCC P-1700	250	59
Polycarbonate, Lexan 113-T	250	90
Polyester, Paraplex P-43	125	100
Polyester, Hetron 92	125	100
Polyurethane rigid foam	250	50
Polyurethane rigid foam, FR	250	24
Polystyrene rigid foam, FR	250	56
Polyvinyl chloride rigid foam	250	51

SECTION 4.9. TESTS FOR SMOKE EVOLUTION

Smoke density may be defined as the degree of light or sight obscuration produced by the smoke from the burning material under given conditions of combustion. This characteristic provides a measure of fire hazard, in that an occupant has a better chance of escaping from a burning structure if he can see the exit, and a firefighter has a better chance of putting out the fire if he can locate it. Some measures of smoke density are degree of light absorption, specific optical density, and smoke development factor.

Tests for smoke evolution generally involve the measurement of the fraction of light absorbed or obstructed by smoke evolved from a decomposing or burning material. The degree of obscuration is a function of the number and size of particles, refractive index, light scattering, rate of movement, extent of ventilation, and distance through which light must travel. The practical effects of smoke can be studied without detailed consideration of particle size and distribution, and the smoke production of various materials can be compared under equal conditions of test.

Tests for smoke-producing characteristics employ one of two measurement techniques: gravimetric and optical. Some tests are based on determining the weight of smoke particles deposited on a filter under specified conditions; other tests involve measurement of the fraction of light absorbed or obstructed by the smoke evolved. The best-known gravimetric method is the Arapahoe smoke test (4.901). The most widely used optical methods are the ASTM E 662 test (4.902), the ASTM D 2843 test (4.903) and the OSU release rate test (4.602). The ASTM E 662 and D 2843 tests measure the density of smoke accumulated in an enclosure, while the ASTM E 84 and OSU tests measure the density of smoke flowing past a specific location.

The Arapahoe smoke chamber (4.901) consists of a vertical cylindrical combustion chamber 125 mm (5 in) in diameter and 175 mm (7 in) high, a cylindrical chamber stack 75 mm (3 in) in diameter and 450 mm (18 in) high, and a filter assembly at the top of the stack. A propane burner mounted at an angle of 10 degrees from the horizontal is mounted in the base of the combustion chamber, and is fed with approximately 90 cm^3/min of propane to produce a well-defined blue flame about 25 mm (1 in) long. A specimen 38 by 13 by 3 mm (1.5 by 0.5 by 0.125 in) is exposed to the burner flame, and the smoke particles are collected on the surface of the glass fiber filter paper at the top of the stack by drawing air through the filter under controlled vacuum. Some data obtained using this test (4.901, 4.904, 4.905) are presented in Table 4.7.

The ASTM E 662 or National Bureau of Standards smoke chamber (4.902) is a completely closed cabinet, 914 by 610 by 914 mm (36 by 24 by 36 in), in which a specimen 76.2 mm (3 in) square is supported vertically in a frame such that an area 65.1 mm (2-9/16 in) square is exposed to heat under either piloted (flaming) or nonpiloted (smoldering) conditions. The heat source is an electric furnace, adjusted with the help of a circular foil radiometer to give a heat flux of 2.5 W/cm^2 (2.2 Btu/sec-ft^2) at the specimen surface. A vertical photometer path for measuring light absorption is employed to minimize measurement differences due to smoke stratification which could occur with a horizontal photometer path at a fixed height; the full 914 mm (3 ft.) height of the chamber is used to provide an overall average for the entire chamber. Some test data are shown in Table 3.6.

The ASTM D 2843 test (4.903) employs a chamber 300 by 300 by 790 mm (12 by 12 by 31 in), completely sealed except for 25 by 230 mm (1 by 9 in) openings of the four sides of the bottom of the chamber. A specimen 25.4 by 25.4 by 6.2 mm (1 by 1 by ¼ in) is exposed to the propane-air flame from a burner with a 0.13 mm (0.005 in) orifice operating at a pressure of 2.8 kgf/cm² (40 psi). A horizontal photometer path, 480 mm (19¾ in) above the base of the chamber, is used for measuring light absorption. Some data obtained using this test (4.909) are shown in Table 4.8.

The ASTM E 84 tunnel test (4.501) requires a specimen 7620 by 496 mm (25 ft. by 19.5 in), mounted face down so as to form the roof of a tunnel 7620 by 445 by 305 mm. The fire source, two gas burners 305 mm (12 in) from the fire end of the sample and 190 mm (7.5 in) below the surface of the sample, is adjusted so that a test sample of select-grade red oak flooring would spread flame 5940 mm (19.5 ft) from the end of the igniting fire in 5.5 min ± 15 sec. The end of the igniting fire is considered as being 1370 mm (4.5 ft) from the burners, this flame length being due to an average air velocity of 73.2 ± 1.5 m/min (240 ± 5 ft./min). A vertical photometer path at the vent pipe 406 mm (16 in) in diameter is used for measuring light absorption. Some ASTM E 84 smoke data are presented in Table 3.4.

The Ohio State State University (OSU) release rate apparatus (4.602) employs a chamber 890 by 410 by 200 mm (35 by 16 by 8 in), with a pyramidal top section 395 mm (15.5 in) high connecting to the outlet. The chamber contains either an electrically heated radiant panel, gas-fired radiant panel, or electrically heated Globar elements. The specimen is positioned in either a vertical or horizontal position for exposure to the heat source, and a horizontal photometer path above the outlet is used for measuring light absorption. The total air flow of 0.04 m³/s (84 ft.³/min) leaves the apparatus through a rectangular exhaust stack 133 by 70 mm (5.25 by 2.75 in) in cross section and 254 mm (10 in) long. Because the horizontal photometer light path is positioned 25 mm (1 in) above the exhaust stack and centered parallel to the length of the opening, the path length is considered to be 133 mm (5.25 in), even though the two parts of the optical system are 180 mm (7 in) apart. Some data obtained using this test are presented in Table 4.9.

It is readily apparent that, under piloted or flaming conditions, many plastics, particularly polyvinyl chloride, polystyrene, and ABS produce more smoke than red oak, based on gravimetric and optical techniques. There were some differences among the different species of wood and among the different types of cellulosic boards, and some plastics which produced much less smoke than polystyrene, but plastics generally seemed to produce more smoke than cellulosics under flaming conditions. Wood, however, produced more smoke under nonpiloted or smoldering conditions than under flaming conditions, with the result that under the former conditions many plastics were comparable to wood in smoke production. Among the plastics, the phenolics generated less smoke than did the olefin and styrene polymers.

The differences between cellulosics and plastics appear to be greatest under flaming conditions with relatively high flow rates. In the ASTM E 84 test, the initial flow rate of 353 ft.^{3}/min produces a velocity of 252 ft./min across the optical path; in the OSU test, the initial flow rate of 84 ft.^{3}/min produces a velocity of 838 ft./min across the optical path. Plastics such as polyvinyl chloride and polystyrene seem to produce smoke in such quantity that the air flow which would dilute smoke from wood has much less dilution effect in the case of these plastics; the use of ventilation with the NBS smoke test demonstrated the difference between these two plastics and wood (4.911).

Table 4.9. Smoke Producing Characteristics of Materials as Measured by the OSU Release Rate Test

Material	Orientation V-vertical h-horizontal	Applied Heat Flux W/cm^2	Maximum Release Rate units/ min-m^3	Total Smoke Release units/m^3	
				3 min	10 min
Oak, 1 in	V	1.0	0.9	0.9	2.3
		2.0	5.5	1.8	4.6
		2.5	6.9	9.1	11.4
Pine, 1 in	V	1.5	2.3	6.9	9.1
		2.5	27.4	32.0	36.6
Red oak, flooring	H	1.0	0.9	0.9	2.3
		2.0	2.3	1.4	4.6
		2.75	6.9	9.1	11.4
Hardboard, 0.25 in	V	1.0	13.7	0	13.7
		2.0	36.6	0.9	73.1

(Continued)

Table 4.9. Smoke-Producing Characteristics of Materials as Measured by the OSU Release Rate Test (Continued)

Material	Orientation V-vertical H-horizontal	Applied Heat Flux W/cm²	Maximum Smoke Release Rate units/ min-m³	Total Smoke Release units/m³ 3 min	10 min
Particle board, 0.5 in	V	1.0	0.9		
		2.5	22.8	9.1	27.4
Particle board, Fr, 0.5 in	V	1.5	0.9		
		2.5	4.6	0.9	4.6
Polyvinyl chloride, flexible, sheet, 90 mil	V	1.0	228	228	640
		2.0	366	640	731
Polystyrene, light diffuser	H	0	77.7	16.0	366
		1.0	137	45.7	457
ABS, sheet, 0.125 in	H	0	274	2.3	137
		1.0	366	155	457
		2.5	366	457	548
Polymethyl methacrylate, light diffuser	H	0	6.9	1.8	9.1
		1.0	4.6	2.3	13.7
Fiberboard, low density, 0.625 in	V	1.0	27.4	45.7	45.7
		2.0	45.7	80.0	80.0
Exterior plywood, 0.5 in	V	1.0	3.4	1.4	4.6
		2.0	4.6	3.7	6.9
Polypropylene, sheet, 0.125 in	H	1.0	228	4.6	686
Polyvinyl chloride, rigid, pipe	V	1.0	4.6	0.9	22.8
		2.6	114	91.4	457
Polysulfone, sheet, 95 mil	H	1.0	0	0	0
		2.5	183	32.0	320
Polyurethane rigid foam, 1 in	V	0	18.3	9.1	9.1
		1.0	27.4	9.1	9.1
Polyurethane rigid foam, FR, 1 in	V	0.5	0.9	0	0.9
		1.0	137	114	114
		2.5	183	137	137
Polyisocyanurate rigid foam, 1 in	V	1.0	9.1	2.3	2.3
		2.75	228	27.4	32.0

(Concluded)

SECTION 4.10. TESTS FOR TOXIC GAS EVOLUTION

Toxic gas evolution may be defined as the level of toxicity exhibited by the gases evolved from a material under specified conditions. This characteristic provides a measure of fire hazard, in that a material which produces more toxic effects than another material would be expected to be more likely to produce incapacitation and death.

Tests for fire gases generally fall into two types: those concerned with identification and analysis of the chemical compounds in the gaseous combustion products, and those concerned with studying the effects of these gases on laboratory animals. Both types of tests suffer from common problems: interference from other compounds, and loss of or change in materials between the point of fire and the point of test. The latter problem, which can present serious difficulty in accounting for the materials in the system, presents an interesting question of validity. If some of the gaseous products have characteristics such that they are condensed on or absorbed by various materials on the way from the combustion chamber to the point of sampling or analysis, eliminating all possibility of condensation or absorption in a laboratory investigation would produce results which would have little relation to the fire gases reaching persons at or near an actual fire. If high temperature is required to keep such wayward compounds in the gas stream, the gas temperatures involved would probably be above the upper limit for survival, and toxicity would be irrelevant to a dead person.

The chemical methods of identification and analysis which may be useful for studying fire gases are:

1. infrared analysis
2. gas chromatographic analysis
3. mass spectrometric analysis
4. colorimetric indicators

If infrared radiation having the same vibration frequency as that characteristic of a molecule passes near the molecule, it will excite the molecule to sympathetic vibrations with consequent loss of radiation energy. When a sample of a material is placed in a beam of infrared radiation, it absorbs energy at those frequencies characteristic of the molecules sent and transmits all other frequencies. By placing a material in an infrared spectrophotometer, the infrared absorption spectrum of the material is obtained and can be compared with the catalogued spectra of known compounds.

In gas chromatography, the material to be analyzed is carried into an adsorbent-containing column by a stream of inert gas. As the flow of carrier gas continues, the components of the material are adsorbed and move down the column, eventually appearing in the effluent gas where each component is detected as it appears.

The mass spectrometer is an apparatus in which the molecules are broken up into charged fragments by electron bombardment. These fragments are deflected in a magnetic field, with the amount of deflection depending on the field intensity and the mass and speed of the fragments. Masses are sorted and measured individually to yield a mass spectrogram, which can be studied for both mass number and relative peak intensities and compared with spectra of known compounds.

Colorimetric indicators are materials which have been calibrated to exhibit a color change when a specific level of a specific compound is reached. They are rapid and convenient to use and are reasonably accurate, but they are subject to error in the event of interference from compounds other than the compounds for which they are calibrated.

Toxicological studies on laboratory animals generally involve exposure of small animals, usually rats or mice, to the gaseous products of decomposition and combustion under carefully controlled conditions, followed by autopsies and examinations to determine cause of death, if fatalities occur, and extent of injuries.

A test procedure for toxicological studies is used in Germany (4.1001) at the hygiene institutes of universities at Hamburg, Munster, and Aix-la-Chapelle, and at the toxicology institutes of Bayer (Wuppertal), BASF (Ludwigshafen), and Hoechst (Frankfurt). The apparatus consists of a horizontal fused silica tube with a length of at least 1000 mm, an outside diameter of 40 mm, and a wall thickness of 2 mm. An annular electric oven tightly enclosing the tube is moved along the tube; its inner diameter must not exceed the outer diamater of the tube by more than 1 mm. The air maintaining the pyrolysis process is introduced into the tube through a flowmeter. The oven moves against the air stream, and the speed of the annular oven determines the burning rate of the material, and, together with the length of the test sample, determines the time of test duration. Sample lengths were 300 and 1000 mm, sample widths were 10, 15, and 25 mm, and sample thicknesses were 2 and 5 mm. Oven speed was 10 or 20 mm per minute. Exposure times were 30 and 60 minutes. Air supply rates were 100 or 300 liters per hour. The animal exposure chambers varied with the laboratory. At the Bayer Institute of Toxicology, 20 male Wistar rats were exposed for 30 minutes in a 50 liter capacity exposure chamber in each test.

These studies were the basis for the development of the German standard DIN 53 436 (4.1002).

A test procedure developed at the University of Tokyo (4.103) exposes a specimen measuring 18 by 18 cm to radiant heat from an electric heater of 1.5 kilowatts and a propane gas flame from a burner fed with 0.35 liter of propane and 3 liters of air per minute. The combustion gases enter a dilution chamber measuring 0.5 by 0.5 by 0.5 meter for adjustment to suitable temperature and concentration, and then led into an exposure chamber measuring 0.5 by 0.5 by 0.5 meter containing 8 mice in movement detecting devices. Four of these devices are of the revolving type (with a freely revolving treadwheel) and four are of the strain type (with a strain gage).

This test procedure was the basis of the Japanese standard test for combustion gas toxicity (4.1004).

A toxicity screening test method was developed under a grant from the National Aeronautics and Space Administration at the University of San Francisco (4.1005–4.1007); the NASA-USF method was further developed by the Product Safety Corporation (4.1008), and has been made a part of the specification for seat cushion material in the Bay Area Rapid Transit District (4.1009). Four Swiss Webster male mice, 25 to 40 g body weight, are placed in a 4.2 liter hemispherical chamber and the off-gases from 1.00 g of material are admitted into the chamber. The test animals are observed for responses such as staggering, convulsions, collapse, and death, and the times to the observation of each response in each animal are recorded. The test is terminated after 30 minutes. The rising temperature program (PSC Procedure 1, or NASA-USF Procedure B) has been used to evaluate over 300 materials under conditions which are intended to simulate the pre-ignition and pre-flashover stages of a fire. Some data obtained using this test method are presented in Table 3.7.

REFERENCES

4.201. "Part 1633. Proposed Standard for the Flammability (Cigarette Ignition Resistance) of Upholstered Furniture (PFF 6-76)," Journal of Consumer Product Flammability, Vol. 4, No. 3, 267–287 (September 1977).

4.202. Damant, G. H., "Flammability Aspects of Upholstered Furniture," Journal of Consumer Product Flamambility, Vol. 3, No. 1, 21–61 (March 1976).

4.203. Damant, G. H., and Young, M. A., "Smoldering Characteristics of Fabrics Used as Upholstered Furniture Coverings," Journal of Consumer Product Flammability, Vol. 4, No. 1, 60–113 (March 1977).

4.204. Hilado, C. J., Brandt, D. L., and Damant, G. H., "Smoldering Tests on Furniture and Aircraft Seat Fabrics," Journal of Consumer Product Flammability, Vol. 5, No. 3, 121–125 (September 1978).

4.301. "Standard Test Method for Behavior of Materials in a Vertical Tube Furnace at 750°C," ANSI/ASTM E 136-79, ASTM Standards, Part 18, 865–867 (1980).

4.302. "Standard Test Method for Ignition Properties of Plastics," ANSI/ASTM D 1929-77, ASTM Standards, Part 35, 652–657 (1980).

4.303. Welker, J. R., Wesson, H. R., and Sliepcevich, C. M., "Ignition of Alpha-Cellulose and Cotton Fabric by Flame Radiation," Fire Technology, Vol. 5, No. 1, 59–66 (February 1969).

4.304. Chien, W. P., and Seader, J. D., "Smoke Development at Different Energy Flux Levels in an NBS Smoke Density Chamber," Fire Technology, Vol. 10, No. 3, 187–196 (August 1974).

4.305. Gaskill, J. R., Taylor, R. D., Ford, H. W., and Miller, H. H., "Development, Calibration, and Use of a High-Flux Heater in the NBS/LLL Smoke Chamber," Journal of Fire and Flammability, Vol. 8, No. 2, 160–179 (April 1977).

4.306. Hilado, C. J., Barnes, G. J., Kourtides, D. A., and Parker, J. A., "The Use of the High Flux Heater in the Smoke Chamber to Measure Ignitability and Smoke Evolution of Composite Panels," Journal of Fire and Flammability, Vol. 9, No. 2, 164–175 (April 1978).

4.307. Hilado, C. J., and Murphy, R. M., "A Simple Laboratory Method for Determining Ignitability of Materials," Journal of Fire and Flammability, Vol. 9, No. 2, 164–175 (April 1978).

4.308. "Standard Method of Testing Rigid Sheet and Plate Materials Used for Electrical Insulation," ANSI/ASTM D 229-77, ASTM Standards, Part 35, 73–94, (1980).

4.309. U.S. General Services Admin., Fed. Test Method Std. No. 406, Method 2023 (October 5, 1961).

4.310. Kline, L. M., "Burning, Arcing, Ignition, and Tracking of Plastics Used in Electrical Appliances," Underwriters Laboratories Research Bulletin No. 55 (February 1964).

4.311. Reymers, H., "New Flammability Indexes, What They Are, What They Mean," Modern Plastics, Vol. 47, No. 10, 92–96, 98 (October 1970).

4.312. U.S. General Services Admin., Fed. Test Method Std. No. 406, Method 4011 (October 5, 1961).

4.401. Manka, M. J., Pierce, H., and Huggett, C., "Studies of the Flash Fire Potential of Aircraft Cabin Interior Materials," FAA-RD-77-47, Federal Aviation Administration, Washington, D.C. (December 1977).

4.402. Trabold, E. L., "Study to Develop Improved Fire Resistance Aircraft Passenger Seat Materials. Phase I," NASA CR-152056, National Aeronautics and Space Administration, Ames Research Center, Moffett Field, California (1977).

4.403. Hilado, C. J., and Cumming, H. J., "Screening Materials for Flash-Fire Propensity," Modern Platics, Vol. 54, No. 11, 56–59 (November 1977).

4.404. Hilado, C. J., and Cumming, H. J., "Flash Fire Propensity of Materials," Journal of Fire and Flammability, Vol. 8, No. 4, 443–457 (October 1977).

4.405. Dorsett, H. G., Jacobson, M., Nagy, J., and Williams, R. P., "Laboratory Equipment and Test Procedures for Evaluating Explosibility of Dusts," U.S. Bureau of Mines Report of Invest. 5624 (1960).

4.501. "Standard Test Method for Surface Burning Characteristics of Building Materials," ANSI/ASTM E 84-80, ASTM Standards, Part 18, 795–808 (1980).

4.502. "Standard Test Method for Surface Flammability of Materials Using a Radiant Heat Energy Source," ANSI/ASTM E 162-79, ASTM Standards, Part 18, 902–913 (1980).

4.503. "Standard Test Method for Critical Radiant Flux of Floor-Covering Systems Using a Radiant Heat Energy Source," ANSI/ASTM E 648-78, ASTM Standards, Part 18, 1232–1248 (1980).

4.504. "Standard Test Method for Surface Flammability of Building Materials Using an 8-ft (2.44-m) Tunnel Furnace," ASTM E 286-75, ASTM Standards, Part 18, 937–942 (1980).

4.505. Hilado, C. J., and Burgess, P. E., "A Four-Foot Tunnel Test Apparatus for Measuring Surface

Flame Spread," Journal of Fire and Flammabilty, Vol. 3, No. 2, 154–163 (April 1972).
4.506. Vandersall, H. L., "The Use of a Small Flame Tunnel for Evaluating Fire Hazard," Journal of Paint Technology, Vol. 39, No. 511, 494–500 (August 1967).
4.507. Wilson, J. A., "Surface Flammability of Materials: A Survey of Test Methods and Comparison of Results," ASTM Spec. Tech. Publ. No. 301, 60–82 (February 1961).
4.508. Levy, M. M., "A Simplified Method for Determining Flame Spread," Fire Technology, Vol. 3, No. 1, 38–46 (February 1967).
4.509. "Standard Test Method for Rate of Burning and/or Extent and Time of Burning of Self-Supporting Plastics in a Horizontal Position," ANSI/ASTM D 635-77, ASTM Standards, Part 35, 218–221 (1980).
4.510. "Standard Test Method for Rate of Burning and/or Extent and Time of Burning of Cellular Plastics Using a Specimen Supported by a Horizontal Screen," ASTM D 1692-76, ASTM Standards, Part 35, 562 (1980).
4.511. Underwriters Laboratories, "Tests for Flammability of Plastic Materials for Parts in Devices and Appliances," Std. UL 94 (February 1, 1974).
4.512. "Standard Test Method for Flame Height, Time of Burning, and Loss of Weight of Rigid Cellular Plastics in a Vertical Position," ASTM D 3014-76, ASTM Standards, Part 35, 775–779 (1980).
4.513. "Standard Test Method for Rate of Burning and/or Extent and Time of Burning of Flexible Plastics in a Vertical Position," ANSI/ASTM D 568-77, ASTM Standards, Part 35, 196–198 (1980).
4.514. "Standard Test Method for Rate of Burning and/or Extent and Time of Burning of Flexible Thin Plastic Sheeting Supporting on a 45-deg. Incline," ANSI/ASTM D 1433-77, ASTM Standards, Part 35, 499–504 (1980).
4.515. "Standard Test Method for Incandescence Resistance of Rigid Plastics in a Horizontal Position," ANSI/ASTM D 757-77, ASTM Standards, Part 35, 307–312 (1980).
4.516. U.S. General Services Admin., Federal Test Method Std. No. 406 (October 5, 1961).
4.517. "Standard Test Method for Fire Retardancy of Paints (Cabinet Method)," ASTM D 1360-79, ASTM Standards, part 27 (1980).
4.518. ASTM D 1361-70, ASTM Standards, Part 27, 212–214 (1974).
4.519. "Standard Test Method for Flammability of Clothing Textiles," ASTM D 1230-72, ASTM Standards, Part 32 (1980).
4.520. "Standard Test Method for Flammability of Finished Textile Floor Covering Materials," ASTM D 2859-76, ASTM Standards, Part 32 (1980).
4.521. U.S. Dept. of Commerce, "Standard for Surface Flammability of Carpets and Rugs," DOC FF 1-70, Federal Register, Vol. 35, No. 74, 6211–6214 (April 16, 1970).
4.522. U.S. Dept. of Commerce, "Standard for Surface Flammability of Small Carpets and Rugs," DOC FF 2-70, Federal Register, Vol. 35, No. 251, 19702–19704 (December 29. 1970).
4.523. U.S. General Services Admin., "Floor Coverings, Resilient, Nontextile," Fed. Test Method Std. No. 501a (June 15, 1966).
4.524. Engerman, H. S., "Floor Covering Systems – A Test for Flame Propagation Index," Journal of Fire and Flammability, Vol. 2, No. 1, 11–35 (January 1971).
4.525. Williamson, R. B., and Baron, F. M., "A Corner Fire Test to Simulate Residential Fires," Journal of Fire and Flammability, Vol. 4, No. 2, 99–105 (April 1973).
4.526. Factory Mutual Research Corp., "Factory Mutual Building Corner Fire Test Procedure" (June 1972).
4.527. "Standard Methods of Fire Tests of Roof Coverings," ANSI/ASTM E 108-80a, ASTM Standards, Part 18, 830–843 (1980).
4.528. "Standard Test Method for Combustible Properties of Treated Wood by the Fire-Tube Apparatus," ANSI/ASTM E 69-80, ASTM Standards, Part 18, 764–769 (1980).
4.529. "Standard Test Method for Combustible Properties of Treated Wood by the Crib Test,"

ANSI/ASTM E 160-80, ASTM Standards, Part 18, 896–901 (1980).
4.530. Underwriters Laboratories, "Flame Tests for Flame-Resistant Fabrics and Films," Std. UL 214 (March 30, 1971).
4.531. U.S. General Services Admin., "Textile Test Methods," Fed. Test Method Std. No. 191 (December 31, 1968).
4.532. U.S. Dept. of Commerce, "Standard for Flammability of Children's Sleepwear," DOC FF 3-71, Federal Register, Vol. 36, No. 146, 14062–14073 (July 29, 1971).
4.533. U.S. Federal Highway Admin., Motor Vehicle Safety Standard No. 302, Federal Register, Vol. 36, No. 5, 289–291 (January 8, 1971).
4.534. "Standard Test Method for Surface Flammability of Flexible Cellular Materials Using a Radiant Heat Energy Source," ASTM D 3675-78, ASTM Standards, Part 38 (1980).
4.535. Nadeau, H. G., and Waszeciak, P. H., "Development of a Small Corner Test Method for the Preliminary Evaluation of Fire Performance of Rigid Cellular Plastics With and Without Protective Covering," Journal of Cellular Plastics, Vol. 12, No. 4, 208 (July/August 1976).
4.601. Factory Mutual Research Corporation, "Fire Hazard Standards, Class I Building Materials" (January 1970).
4.602. "Proposed Test Method for Heat and Visible Smoke Release Rates for Materials," ASTM Standards, Part 18, 1382–1400 (1980).
4.603. Bellisson, G., "The Crucible Test Method for Defining Noncombustible Materials," ASTM Special Technical Publication 502, 83–98 (1972).
4.701. Standard Methods of Fire Tests of Building Construction and Materials," ANSI/ASTM E 119-80, ASTM Standards, Part 18, 844–864 (1980).
4.702. "Standard Methods of Fire Tests of Door Assemblies," ANSI/ASTM E 152-80, ASTM Standards, Part 18, 880–889 (1980).
4.703. "Standard Methods of Fire Tests of Window Assemblies," ANSI/ASTM E 163-80, ASTM Standards, Part 18, 914–919 (1980).
4.704. Way, D. H., and Hilado, C. J., "The Performance of Rigid Cellular Plastics in Fire Tests for Industrial Insulation," Journal of Cellular Plastics, Vol. 4, No. 6, 221–228 (June 1968).
4.705. Hilado, C. J., and Way, D. H., "Fire Performance of Spray-Applied Rigid Urethane Foam as Vessel Insulation," Journal of Fire and Flammability, Vol. 1, No. 1, 30–35 (January 1970).
4.706. Factory Mutual Research Corp., "Test Procedure for Susceptibility to Radiant Heat Damage of Cellular Plastics in Wall and Roof Constructions" (July 1971).
4.707. Underwriters Laboratories, "Air Ducts," Std. UL 181 (May 15, 1972).
4.708. U.S. General Services Admin., Federal Spec. SS-A-118b (April 1959).
4.709. Mitchell, D. W., Murphy, E. M., and Nagy, J., "Fire Hazard of Urethane Foam in Mines," U.S. Bureau of Mines Report of Invest. No. 6837 (1966).
4.710. Mitchell, S. W., Murphy, E. M., and Nagy, J., "Rigid Foam for Mines," U.S. Bureau of Mines Report of Invest. No. 6366 (1964).
4.711. Riccitiello, S. R., Fish, R. H., Parker, J. A., and Gustafson, E. J., "Development and Evaluation of Modified Polyisocyanurate Foams for Low-Heating-Rate Thermal Protection," Journal of Cellular Plastics, Vol. 7, No. 2, 91–96 (March/April 1971).
4.712. Ball, G. W., Haggis, J. A., Hurd, R., and Wood, J. F., "A New Heat Resistant Rigid Foam," Journal of Cellular Plastics, Vol. 4, No. 7, 248–261 (July 1968).
4.801. "Standard Method for Measuring the MInimum Oxygen Concentration to Support Candle-Like Combustion of Plastics (Oxygen Index)," ANSI/ASTM D 2863-77, ASTM Standards, Part 35, 752–758 (1980).
4.901. Hilado, C. J., and Cumming, H. J., "Studies with the Arapahoe Smoke Chamber, Journal of Fire and Flammability, Vol. 8, No. 3, 300–308 (July 1977).
4.902. "Standard Test Method for Specific Optical Density of Smoke Generated by Solid Materials," ANSI/ASTM E 662-79, ASTM Standards, Part 18, 1266–1294 (1980).
4.903. "Standard Test Method for Density of Smoke from the Burning or Decomposition of

Plastics," ANSI/ASTM D 2843-77, ASTM Standards, Part 35, 736–744 (1980).
4.904. Hilado, C. J., and Machado, A. M., "Smoke Studies with the Arapahoe Chamber," Journal of Fire and Flammability, Vol. 9, No. 2, 240–244 (April 1978).
4.905. Hilado, C. J., Machado, A. M., and Murphy, R. M., "Smoke Density Studies with the Arapahoe and NBS Chambers," Journal of Fire and Flammability, Vol. 9, No. 4, 421–425 (October 1978).
4.906. Gross, D., Loftus, J. J., and Robertson, A. F., "Method for Measuring Smoke from Building Materials," ASTM STP 422, 166–204 (1966).
4.907. Gaskill, J. R., and Veith, C. R., "Opacity of Smokes from Certain Woods and Plastics under Smoldering and Burning Conditions," ACS Org. Chem. Plastics Chem. Papers, Vol. 28, No. 1, 282–290 (April 1968).
4.908. Brenden, J. J., "Determining the Utility of a New Optical Test Procedure for Measuring Smoke from Various Wood Products," FPL 137, Forest Products Laboratory, Madison, Wisconsin (June 1970).
4.909. Hilado, C. J., "The Effect of Chemical and Physical Factors on Smoke Evolution from Polymers," Journal of Fire and Flammability, Vol. 1, No. 3, 217–238 (July 1970).
4.910. Nelson, G. L., "Smoke Evolution: Thermoplastics," Journal of Fire and Flammability, Vol. 5, No. 2, 125–135 (April 1974).
4.911. Gaskill, J. R., "Smoke Development in Polymers During Pyrolysis or Combustion," Journal of Fire and Flammablity, Vol. 1, No. 3, 183–216 (July 1970).
4.1001. Kimmerle, G., "Aspects and Methodology for the Evaluation of Toxicological Parameters during Fire Exposure," Journal of Fire and Flammability, Combustion Toxicology, Vol. 1, No. 1, 4–51 (February 1974).
4.1002. Klimisch, H.-J, Hollander, H. W. M., and Thyssen, J., "Comparative Measurements of the Toxicity to Laboratory Animals of Products of Thermal Decomposition Generated by the Method of DIN 53 436," Journal of Combustion Toxicology, Vol. 7, No. 4, 209–230 (November 1980).
4.1003. Kishitani, K., and Nakamura, K., "Toxicities of Combustion Products," Journal of Fire and Flammability, Combustion Toxicology, Vol. 1, No. 2, 104–120 (May 1974).
4.1004. Tsuchiya, Y., "New Japanese Standard Test for Combustion Gas Toxicity," Journal of Combustion Toxicology, Vol. 4, No. 1, 5–7 (February 1977).
4.1005. Hilado, C. J., and Cumming, H. J., "Relative Toxicity of Pyrolysis Gases from Materials: Effects of Chemical Composition and Test Conditions," Fire and Materials, Vol. 2, No. 2, 68–79 (April 1978).
4.1006. Hilado, C. J., Cumming, H. J., and Machado, A. M., "Relative Toxicity of Pyrolysis Gases from Materials: Specific Toxicants and Special Studies," Fire and Materials, Vol. 2, No. 4, 141–153 (October 1978).
4.1007. Hilado, C. J., Cumming, H. J., and Schneider, J. E., "Relative Toxicity of Pyrolysis Gases from Materials: Effects of Temperature, Air Flow, and Criteria," Fire and Materials, Vol. 3, No. 4, 183–187 (December 1979).
4.1008. Hilado, C. J., and Huttlinger, P. A., "Concentration-Time Relationships in Off-Gas Toxicity," Fire Technology, Vol. 16, No. 3, 192–203 (August 1980).
4.1009. Jenkins, C. E., and Procter, H. S., "BART Seat Specification Development," Proceedings of the California Conference on Fire Toxicity, Vol. 2, 80–109 (1980).

CHAPTER 5

Prevention, Inhibition, and Extinguishment

The work of preventing, inhibiting, and extinguishing fire can be divided into two areas: design of the material, and design of the application.

SECTION 5.1. DESIGN OF THE MATERIAL

Flame retardance involves the disruption of the burning process at one or more stages so that the process is terminated within an acceptable period of time, preferably before ignition actually occurs. When dealing with plastic materials, the following general approaches are available:

1. Design of the basic polymer so that exposure to heat and oxygen will not produce combustion. This is achieved by producing thermally stable polymers with high decomposition temperatures and high fractions of solid residue after decomposition. Such polymers are being developed, but they presently lack the versatility in processing and performance characteristics needed for wide acceptance and more favorable economics.

2. Modification of existing polymers so that they exhibit satisfactory performance upon exposure to fire. Examples of this approach are the chlorination of polyethylene, the substitution of chlorendic or tetrabromophthalic anhydride for other compounds in polyesters, and the use of phosphorus-containing polyols in polyurethanes. These substi-

tuted materials are known as reactive type flame retardants.

3. Incorporation of compounds so that the resulting plastic materials will exhibit satisfactory performance upon exposure to fire. Examples of this approach are the addition of tris (2,3-dibromopropyl) phosphate, and the addition of barium metaborate. These added materials are known as additive type flame retardants.

The first approach provides little help in imparting flame retardance to the materials being presently used at the rate of several billion pounds annually, and it is doubtful that the new high-temperature polymers will ever acquire sufficient versatility in processing and performance and sufficiently favorable economics to displace presently used materials to a significant extent.

The third approach provides the most expeditious means of providing flame retardance, particularly at the end-use or compounding stage. The incorporation of additives, however, has limitations and disadvantages, and the second approach is receiving increasing attention because fire resistance is simply added to the list of properties considered desirable when modifying existing polymers.

This section will therefore be concerned with the use of additive type and reactive type flame retardants.

The general approaches involved in the use of additive type flame retardants are the following:

1. Incorporation of a compound that will redirect the decomposition and combustion reactions toward the evolution of gases which are noncombustible, or heavy enough to interfere with normal interchange of combustion gases and combustion air. Ammonium carbonate decomposes into ammonia, water, and carbon dioxide, and tetrabromoethane decomposes to evolve hydrogen bromide. Phosphorus compounds are believed to alter the decomposition reactions so as to increase the amount of carbonaceous residue at the expense of flammable gases. Halogen compounds are believed to inhibit the free radical chain reactions involved in decomposing the polymer into combustible gases. This approach essentially constitutes inerference with Stages II and III of the burning process on the macro scale (Decomposition and Ignition), with effects on Stages IV and V (Combustion and Propagation).

2. Incorporation of a compound that will redirect the decomposition and combustion reactions toward reduced heat of combustion. Phosphorus compounds appear effective in this approach, in favoring the formation of carbonaceous char rather than carbon-containing flammable gases, and in inhibiting glowing oxidation of the char. One gram-

mol of carbon produces 26,400 calories of heat in burning to carbon monoxide and 94,000 calories of heat in burning to carbon dioxide; this heat is not produced if carbon is kept in the form of char. This approach essentially constitutes interference with Stage IV of the burning process on the macro scale (Combustion), with effects on Stage V (Propagation).

3. Incorporation of a compound that will conserve the physical integrity of the material, in order to impede access of oxygen and heat, and reduce disintegration of the structure. Fillers and reinforcing glass fibers contribute to physical integrity by assuring certain minimum levels of solid residue. Phosphorus compounds appear effective in this approach in favoring retention of the carbon as char, since carbon monoxide and carbon dioxide are gases and contribute nothing to physical integrity. This approach essentially constitutes interference with Stage II of the burning process on the macro scale (Decomposition), with effects on Stages III, IV, and V (Ignition, Combustion, and Propagation).

4. Incorporation of a compound that will increase the specific heat or thermal conductivity of a material, or otherwise increase dissipation or absorption of heat, to reduce the heat available for propagation. High-density fillers tend to increase heat absorption and dissipation, and inorganic hydrates such as hydrated aluminum oxide require energy for dehydration. This approach essentially constitutes interference with Stage I of the burning process on the macro scale, with effects on succeeding stages. It is, however, not feasible for plastic materials whose primary function is thermal insulation, such as rigid cellular plastics.

The general approaches involved in the use of reactive type flame retardants are the following:

1. Modification of the polymer to favor decomposition and combustion reactions which are non-combustible, or heavy enough to interfere with normal interchange of combustion gases and combustion air. Phosphorus is introduced into polyurethane polymers in the form of phosphonate, phosphate, and phosphite polyols. Halogens are introduced into polyester polymers in the form of halogenated anhydrides such as chlorendic anhydride and tetrabromophthalic anhydride, into epoxy polymers in the form of halogenated compounds such as tetrabromobisphenol-A, and into polyolefins in the form of chlorinated paraffins and olefins.

2. Modification of the polymer to favor decomposition and combustion reactions involving reduced heat of combustion. Phosphorus compounds appear effective in this approach.

3. Modification of the polymer to increase the amount of solid

residue, to maintain structural integrity and to impede access of oxygen and heat. Phosphorus compounds are effective in this approach.

Char formation in polyurethanes is improved by substituting polymeric polyaryl isocyanates for tolylene diisocyanate, and by employing sucrose-based polyethers.

4. Modification of the polymer to reduce ease of ignition, by increasing minimum ignition temperature or decomposition temperature, or by increasing energy required for decomposition. Phosphorus and halogen compounds appear effective in this approach.

The promotion of char formation is perhaps the most important overall concept in flame-retarding plastic materials, because retention of the carbon in the char has the following important benefits:

1. Reduction of heat of combustion, from 94,000 calories per gram mol of carbon dioxide, to 26,400 calories per gram mol of carbon monoxide, to nil for solid carbon.
2. Conservation of mechanical integrity, since retained carbon is a solid.
3. Reduction of oxygen depletion.
4. Reduction of toxicity in fire gases.

The reactive type flame retardants find their greatest application in thermosetting plastic materials, since substitution of compounds can be effected at a stage much closer to end-use than is the case with thermoplastics. Thermoplastics appear to use additive type flame retardants in the majority of cases.

SECTION 5.2. FLAME RETARDING VARIOUS POLYMERS

Any of the approaches discussed theoretically could be used to improve the flame retardance of any polymer. Not all approaches, however, are equally effective from the performance viewpoint, and some are clearly more desirable than others from the economic viewpoint. Some polymers are inherently flame-retardant and require no further improvement for most applications. Other polymers are now used in applications in which fire behavior is not presently a factor. The polymers for which flame retardant technology has been sufficiently developed to warrant some generalizations are therefore those used in large amounts for applications requiring flame retardance.

The discussion in Chapter 1 tends to indicate which types of polymers have been the subject of flame retardant technology: polyethylene, polypropylene, vinyl chloride polymers and copolymers, polystyrene and its copolymers, cellulosics, polyesters and alkyds, and polyurethane foams. Polymers possessing high temperature stability, such as polytetrafluoroethylene and polyimide, will probably never need further flame retardance.

Polyethylene is combustible and requires modification for flame retardance. The approaches used are:

1. Halogenation of the polymer, as in the case of chlorinated polyethylene.

2. Addition of halogen-containing compounds, such as chlorinated paraffins. Aliphatic chlorine compounds are more effective than aromatic chlorine compounds.

3. Use of antimony oxide to increase effectiveness of halogen.

4. Use of phosphorus to increase effectiveness of halogen. Phosphorus used alone requires high concentrations, and phosphorus/bromine is more effective than phosphorus/chlorine.

Polypropylene is combustible and requires modification for flame retardance. The approaches used are:

1. Addition of halogen-containing compounds such as chlorinated paraffins. Aliphatic chlorine compounds are more effective than aromatic chlorine compounds.

2. Use of antimony oxide to increase effectiveness of halogen.

3. Use of phosphorus to increase effectiveness of halogen. Phosphorus used alone requires high concentrations, and phosphorus/bromine is more effective than phosphorus/chlorine.

Polyvinyl chloride as a pure polymer is inherently flame retardant because of its chlorine content of 56 percent. Many applications, however, require the use of plasticizers to impart the desired degree of flexibility. Many plasticizers, such as phthalates, increase flammability by diluting the high-chlorine polymer with combustible material. The approaches used are:

1. Development of phosphate plasticizers to replace phthalate and other plasticizers without loss of plasticizing action, low temperature embrittlement, or reduction in light resistance.

2. Use of secondary plasticizers such as halogenated paraffins, biphenyls, and fatty acid esters.

3. Use of antimony oxide in plasticized formulations.

Polyvinyl acetate, polyvinyl alcohol, and **polyvinyl butyral** are combustible and require modification for flame retardance. The approaches used are:

1. Use of phosphate plasticizers.

2. Use of secondary plasticizers such as halogenated paraffins.

3. Copolymerization with phosphorus or halogen-containing monomers.

Polystyrene is combustible and requires modification for flame retardance. The approaches used are:

1. Addition of halogenated compounds. Bromine is more effective than chlorine, and aliphatic and alicyclic bromine compounds are more effective than aromatic bromine compounds.

2. Chemical bonding of halogenated compounds into the polystyrene chain.

3. Improvement of efficiency of halogens by using small amounts of free radical initiators such as peroxides.

Acrylonitrile-butadiene-styrene (ABS) is combustible and requires modification for flame retardance. The approaches used are:

1. Addition of halogenated compounds.

2. Terpolymerization with halogen-containing monomers, such as bis(2,3-dibromopropyl) fumarate.

Polymethyl methacrylate is combustible and requires modification for flame retardance. The approaches used are:

1. Addition of compounds containing phosphorus and halogen.

2. Combination with polyvinyl chloride to form an alloy.

Cellulose nitrate is extremely flammable and its reputation as a fire hazard has contributed to widespread suspicion of plastic materials in general. The approaches used are:

1. Complete elimination of cellulose nitrate, largely by replacement with less flammable cellulose acetate.

2. Use of phosphate and halogenated phosphate plasticizers.

Cellulose acetate, cellulose butyrate and **cellulose propionate** are combustible and require modification for flame retardance. The approaches used are:

1. Use of phosphate and halogenated phosphate plasticizers.

2. Addition of phosphorus and halogen containing compounds.

Polyesters are combustible unless formulated to be flame retardant. The approaches used are:

1. Use of relatively large amounts of inorganic fillers.

2. Use of flame retardant plasticizers such as halogenated paraffins and phosphates.

3. Substitution of intermediates that contain halogen, such as tetrachlorophthalic anhydride and tetrabromophthalic anhydride.

4. Substitution of intermediates that are both halogen-containing and

more temperature-resistant, such as chlorendic anhydride.

Polyurethanes are combustible unless formulated to be flame retardant. The approaches used are:

1. Use of phosphate plasticizers, where flexibility is desired.
2. Use of polymeric diphenylmethane type isocyanates instead of toluene diisocyanate.
3. Introduction of phosphorus-containing intermediates, such as phosphate, phosphite, phosphonate, and amino-phosphonate polyols.
4. Introduction of halogen-containing intermediates, such as brominated isocyanates and chlorine-containing polymer/polyols.
5. Use of chlorine and fluorine-containing compounds as blowing agents in foams.
6. Use of relatively large amounts of inorganic fillers.

Epoxy resins are combustible unless formulated to be flame retardant. The approaches used are:

1. Introduction of halogen into the bisphenol component, in the form of tetrachlorobisphenols and tetrabromobisphenols.
2. Use of halogenated anhydrides as hardeners, such as chlorendic anhydride.
3. Use of relatively large amounts of inorganic fillers.

SECTION 5.3. FLAME RETARDANCE MECHANISMS

Much of the development of flame retardants has been based on empirical work, because the mechanisms of flame extinguishment are not completely understood. On the basis of empirical results, various theories have been postulated to describe the mechanisms involved in flame inhibition and extinguishment.

The compounds that have been found to be most effective in producing flame retardance are compounds containing phosphorus, bromine, or chlorine, or two or more of these elements. Other elements which have exhibited some flame retardant effect are antimony, arsenic, boron, and nitrogen. Because phosphorus, bromine, and chlorine have proven so much more effective than the other elements cited, considerable work has gone into study of the possible mechanisms involved in their inhibiting and extinguishing action.

In the discussion of the mechanisms involved in the performance of various elements considered to be flame retardants, four concepts should be borne in mind:

1. The phase in which retardance occurs (gaseous or condensed) and

the stage of the burning process at which retardance occurs are important.

2. The flame retardant must be available and active at the right time and in the right place.

3. To impart a specific level of flame retardance to a specific material using a specific flame retardant, a specific amount of that flame retardant is required, and amounts exceeding the optimum are unnecessary and sometimes undesirable.

4. A flame retardant is undesirable for a given material if the amount required has an intolerable effect on other properties.

Flame retardant action is attributed to a variety of compounds which contain elements in two general groups of the Periodic Table of the Elements: Group V, consisting of nitrogen, phosphorus, arsenic, antimony, bismuth, and heavier elements; and Group VII, consisting of fluorine, chlorine, bromine, iodine, and heaver elements. Another element known to have flame retardant value is boron.

The three elements generally considered to provide the most effective flame retardance will be the first discussed; phosphorus, bromine, and chlorine.

Phosphorus is believed to perform most of its flame retardant function in the condensed phase (including both the solid and liquid phases, since varying degrees of melting are involved at fire temperatures). Phosphorus-containing compounds increase the amount of carbonaceous residue or char formed, by one or both of two mechanisms: redirection of the chemical reactions involved in decomposition in favor of reactions yielding carbon rather than carbon monoxide or carbon dioxide; and formation of a surface layer of protective char which inhibits access of oxygen and escape of oxides of carbon by physical blockade, and thus prevent gasification of the carbon by simply denying the oxygen required.

The effect of phosphorus on the chemical reactions involved in decomposition may involve partial oxidation followed by dehydration to yield water, a noncombustible gas with relatively low heat of formation, and carbon.

The effect of phosphorus on the surface characteristics of the char may involve the formation of nonvolatile oxides of phosphorus which act as a flux for the carbonaceous residue.

The action of phosphorus-containing compounds is particularly desirable for the following reasons:

1. It is effective in Stage II (Decomposition) rather than in the later

Stages III (Ignition) and IV (Combustion), increasing the probability of arresting the burning process at an earlier stage.

2. It is effective in the condensed phase, which is the fraction of the material which may have some residual value, rather than in the gaseous phase, which contains the fraction of the material which has no further practical value.

3. It interferes with the oxidation reactions which are responsible for the one major source of heat in the burning of plastic materials: carbon oxidation.

4. It interferes with gasification, which is largely responsible for increased heat transfer and flame propagation.

For greatest effectiveness, a flame retardant should be made available and exhibit its greatest activity at the precise conditions of time and space at which the burning process is most vulnerable to its particular mechanism of extinguishment. A flame retardant that is made available and active at optimum conditions will prove significantly more effective than a similar, supposedly flame retardant, compound which becomes available and active at a less propitious time or in a less desirable place.

Phosphorus content alone is not an adequate measure of the effectiveness of flame retardants. The occasionally reported cases where a phosphorus compound appears less effective than another despite its higher phosphorus level are manifestations of this phenomenon.

A particular combination of fire conditions and material properties requires a specific amount of flame retardant action to provide a specific level of flame retardance. Once the necessary amount of flame retardant action is provided, any increase becomes superfluous and in some cases may be undesirable.

For this reason, increasing phosphorus content provides an increasing degree of flame retardance up to a certain optimum level, beyond which further increases in phosphorus content provide no further benefit. For polyethylene and polypropylene containing no other flame retardants, self-extinguishing behavior requires about 15 percent phosphorus. For vinyl polymers containing no other flame retardants, such as polyvinyl alcohol, self-extinguishing behavior requires about 15 percent phosphorus. For cellulose containing no other flame retardants, in the form of paper or cotton, self-extinguishing behavior requires 3 to 4 percent phosphorus. For polyesters containing no other flame retardants, self-extinguishing behavior requires 2 to 8 percent phosphorus. For polyurethanes containing no other flame retardants, self-extinguishing behavior requires from 1.0 to 1.5 percent phosphorus. Among the polyurethanes, the higher phosphorus requirements of flexible foams for self-extinguishing behavior (6 to 8 percent) can be attributed to large surface area exposed by the interconnected cell structure.

It is noteworthy that the polymers which tend to burn without formation of char (polyolefins and vinyls) require considerably more phosphorus for self-extinguishing behavior (about 15 percent) than the polymers which tend to form on burning (cellulose, polyesters, polyurethanes). Apparently that much additional phosphorus is required to provide sufficient phosphorus pentoxide (the most probable oxide form) for the dehydrating action required to promote char and the fluxing action required to stabilize it.

Many phosphorus esters are susceptible to hydrolysis. For this reason, phosphorus is used in pendant groups rather than in the main polymer chain in polyesters, and its use in polyurethanes is limited by humid aging behavior.

Bromine is believed to perform most of its flame retardant function in the gaseous phase, by means of two mechanisms: redirection or termination of the chemcial reactions involved in combustion, and the evolution of heavy bromine-containing gases which tend to protect the condensed phase by inhibiting access of oxygen and transfer of heat.

The chemical reactions may involve halogenation followed by dehydrohalogenation, to yield a polymeric residue rich in double bonds which is then converted to carbon.

The action of bromine-containing compounds appears to be principally effective in Stage III (ignition), by blocking access of oxygen and heat, and in Stage IV (Combustion), by influencing the combustion reactions. It is necessary for the burning process to proceed through Stage II (Decomposition) and release bromine-containing compounds into the gaseous phase for flame retardant action to occur.

Aliphatic and alicylic bromine compounds have been found to be more effective than aromatic bromine compounds in flame retarding polystyrene. This difference is attributed to the higher bond energy of aromatic bromine, which makes it less readily available for flame retardant action.

For polyesters containing no other flame retardants, self-extinguishing behavior requires about 17 percent bromine.

Comparisons of relative effectiveness of phosphorus and bromine indicate that 1 percent of phosphorus is equivalent to about 6 percent of bromine in polyolefins and polystyrene, about 4 to 5 percent of bromine in polyesters, and about 5 to 7 percent of bromine in polyurethanes.

The concentrations of bromine that can be employed are limited in various polymers by its effect on flexibility, mechanical properties, heat distortion, or durability.

Chlorine is believed to perform its flame retardant function in both the gaseous and condensed phases. In the gaseous phase, it employs the chemical mechanism of redirection or termination of the chemical reactions involved in combustion, and the physical mechanism of evolution of heavy chlorine-containing gases to protect the condensed phase by inhibiting access of oxygen and heat. In the condensed phase, it redirects the chemical reactions involved in decomposition.

The chemical reactions may involve halogenation followed by dehydrohalogenation, to yield a polymeric residue rich in double bonds which is then converted to carbon.

While chlorine-containing compounds would, like bromine, appear to be primarily effective on Stages III (Ignition) and IV (Combustion), its influence on decomposition indicates some effectiveness in Stage II (Decomposition).

Aliphatic chlorine is more effective than aromatic chlorine in both polyesters and polyurethanes. The form in which chlorine is available influences its effectiveness. In polypropylene, perchlorofulvalene is more efficient than the perchloropentacyclodecane, perhaps because perchlorofulvalene decomposes above 250°C. into hexachlorocyclopentadiene and hydrogen chloride.

For epoxy resins containing no other flame retardants, self-extinguishing behavior requires about 16 to 20 percent chlorine. For polyesters containing no other flame retardants, self-extinguishing behavior requires about 21 percent alicyclic chlorine or 28 to 31 percent aromatic chlorine (in forms such as hexahalobenzenes and chlorinated biphenyls).

Large quantitites of halogen can produce poor weathering resistance in polyesters. Highly chlorinated paraffins are preferred over lower chlorinated waxes because of the latter's plasticizing effect. If tetrachlorophthalic anhydride, which contains only 49.5 percent chlorine, is employed in sufficient quantity to bring chlorine content in the polyester to the 30 percent level, the resulting polyester is inferior in heat distortion and mechanical properties.

Phosphorus-bromine synergism is the term applied to the observed ability of combinations of phosphorus and bromine containing compounds to provide a greater degree of flame retardance than equal amounts of each used separately. This phenomenon is utilized to reduce the amounts of any of the two required. In polyolefins, for example, the use of phosphorus alone is not feasible because of the high levels required, and bromine/phosphorus combinations are effective.

The synergism can be explained by the mechanisms employed by phosphorus and bromine compounds. Without phosphorus, hydrogen bromide is believed to be the most active bromine compound. Since this compound boils at −67 °C., only its heavy molecular weight (80.92) is responsible for its relatively long residence in the combustion zone, where it participates in successive halogenation and dehydrohalogenation. The presence of phosphorus promotes the formation of char, which further restricts movement in the gaseous phase, and results in the formation of phosphorus tribromide, phosphorus pentabromide, and phosphorus oxybromide, which are less readily gasified and are heavier gases (molecualr weights of 270.70, 430.52, and 286.70, respectively). Since these bromides are continually regenerated in the halogenation-dehydrohalogenation process, their effectiveness increases with their residence time in the combustion region, and keeping them in effective contact twice as long would tend to make them twice as effective.

The fluxing action of phosphorus is probably limited to the interface of the combustion and pyrolysis regions, so that any excess over the amount needed to stabilize the char is ineffective, and this excess is available for the formation of bromides without any loss of fluxing action. Since phosphorus oxides can effect only so much dehydration while the continually regenerated bromides can repeat the dehydrohalogenation process, phosphorus/bromine combinations would be more effective than phosphorus oxides in promoting char formation.

Phosphorus-chlorine synergism is similar to phosphorus-bromine synergism. It is more extensively utilized because of the greater variety of available chlorine-containing compounds, but it appears to be less effective, to the same degree and for the same reasons that chlorine is not as effective on an equal weight basis as bromine. The chlorides of phosphorus are lower-boiling and lighter gases than the corresponding phosphorus bromides, and can be expected to have a shorter residence time in the combustion zone.

The effectiveness of bromine and chlorine as flame retardants brings attention to the other two well-known members of the halogen family, fluorine and iodine. As molecular weight increases progressively from fluorine through chlorine and bromine to iodine, density of analogous gaseous compounds increases, chemical activity decreases, and chemical bond strength decreases.

Fluorine can be considered a flame retardant element in that its bond to carbon is so strong that it is difficult to render the attached carbon available for oxidation. Where the hydrogen atoms in the polymer have been completely replaced with fluorine atoms, as in the case of polytetrafluoroethylene, decomposition temperatures are high and the decomposition products are noncombustible. Fluorine therefore has a certain amount of effectiveness in Stage II (Decomposition) and Stage III (Ignition). Fluorine-containing gases, however, are lighter than their bromne and chlorine-containing analogs, so that their smothering or blanketing effect is nil by comparison.

Iodine has such a weak bond to carbon and to organic compounds in general that it is likely to initiate decomposition in normal service, and its relative inactivity does not make it likely to influence combustion reactions.

Just as the Group VII elements (fluorine, chlorine, bromine, and iodine) have similarities in flame retardant action, the elements in Group V (nitrogen, phosphorus, arsenic, antimony, and bismuth) also behave as flame retardants with some similarities.

Arsenic is known to have some flame retardant action, but its toxicity has essentially eliminated it from consideration for general use.

Antimony appears ineffective when used by itself. Antimony trioxide melts at 1550°C., a temperature too high to render it an effective fluxing agent in the manner of phosphorus, and it does not have the dehydrating action of phosphorus pentoxide. It is principally used as a synergist with the halogens, particularly bromine and chlorine. Where polyesters containing no other flame retardant require about 17 percent bromine for self-extinguishing behavior, similar results can be obtained with 13 percent bromine and 1 percent antimony trioxide, or 9 percent bromine and 2 percent antimony trioxide. The increased effectiveness is probably due to the longer residence time of bromide as antimony tribromide, relative to bromide as hydrogen bromide. The greater effectiveness of bromine compared to chlorine is also observed here. For polyesters containing 5 percent antimony trioxide, 6 percent bromine produces the same level of flame retardance as 24 to 25 percent chlorine.

Bismuth is known to have some flame retardant action, but it is less active and less well known than antimony.

Nitrogen is known to have some flame retardant action, particularly as a synergist with phosphorus, Phosphorus-nitrogen synergism has been found effective with cellulose, and may be a factor in polyesters and polyurethanes. Acrylonitrile and triethyl phosphate exhibited some

Table 5.1. Properties of Possible Flame Retardant Compounds

Material	Formula	Mol. Wt.	Melting Point °C.	Boiling Point °C.	Decomp. Temp. °C.
Phosphorus trioxide	P_2O_3	109.95	23.8	173.8	
Phosphorus tetroxide	P_2O_4	125.95	>100		
Phosphorus pentoxide	P_2O_5	141.94	580-585	300(s)	
Phosphorus trifluoride	PF_3	87.97	−151.5	−101.5	
Phosphorus pentafluoride	PF_5	125.97	83	75	
Phosphorus trichloride	PCl_3	137.33	−112	75.5	
Phosphorus pentachloride	PCl_5	208.24		162(s)	166.8
Phosphorus tribromide	PBr_3	270.70	− 40	172.9	
Phosphorus pentabromide	PBr_5	430.52			106
Phosphorus oxychloride	$POCl_3$	153.33	2	105.3	
Phosphorus oxybromide	$POBr_3$	286.70	56	189.5	
Hydrogen fluoride	HF	20.01	− 83.1	19.54	
Hydrogen chloride	HCl	36.46	−114.8	− 84.9	
Hydrogen bromide	HBr	80.92	− 88.5	− 67.0	
Hydrogen iodide	HI	127.91	− 50.8	− 35.38	
Antimony trioxide	Sb_2O_3	291.50	656	1550	
Antimony trifluoride	SbF_3	178.75	292	319(s)	
Antimony pentafluoride	SbF_5	216.74	7	149.5	
Antimony trichloride	$SbCl_3$	228.11	73.4	283	
Antimony pentachloride	$SbCl_5$	299.02	28	79	
Antimony tribromide	$SbBr_3$	361.48	96.6	280	
Antimony oxychloride	SbOCl	173.20			170
Arsenic trioxide	As_2O_3	197.84	315		
Arsenic pentoxide	As_2O_5	229.84			315
Arsenic trifluoride	AsF_3	131.92	− 8.5	− 63	
Arsenic trichloride	$AsCl_3$	181.28	− 8.5	63	
Arsenic tribromide	$AsBr_3$	314.65	32.8	221	
Arsenic triiodide	AsI_3	455.64	146	403	
Boron trioxide	B_2O_3	69.62	460	1860	
Boron trifluoride	BF_3	67.81	−126.7	− 99.9	
Boron trichloride	BCl_3	117.17	−107.3	12.5	
Boron tribromide	BBr_3	250.54	− 46	91.3	
Boron triiodide	BI_3	391.52	49.9	210	
Boron carbide	B_4C	55.26	2350	>3500	

synergism in polyesters, and amino-phosphonate polyols are employed in polyurethanes. Phosphorus-nitrogen combinations, however, have been found ineffective in polyolefins.

Boron is believed to perform most of its flame retardant function in the condensed phase. Boron-containing compounds increase the amount of char formed, by one or both of two mechanisms: redirection of the chemical reactions involved in decomposition in favor of reactions yielding carbon rather than carbon monoxide or carbon dioxide; and formation of a surface layer of protective char which prevents gasification of the carbon by blocking access of the oxygen required.

The chemical influence of boron may involve removal of vulnerable hydroxyl groups by dehydration, causing increased formation of char. The physical influence of boron may involve the formation of nonvolatile boric oxide which acts as a flux for the carbonaceous residue.

The action of boron-containing compounds is desirable because:

1. It is effective in Stage II (Decomposition) rather than in later stages, increasing the probability of arresting the burning process at an earlier stage.

2. It is effective in the condensed phase, which is the fraction of the material which may have some residual value.

3. It interferes with carbon oxidation and gasification.

Hydrated boron compounds, such as boric acid, appear to be generally more effective than anhydrous boric oxide at equal boron concentration. This is believed due to the removal of heat required to effect dehydration.

SECTION 5.4. SMOKE RETARDING VARIOUS POLYMERS

Theoretically, the ideal method of flame retarding polymers would also result in a reduction in smoke. This has been true in some situations, particularly those in which increased retention of material in the solid phase or char has resulted in less combustible compounds and less smoke-producing compounds in the gas phase.

In some cases, however, smoke production is greater in the preignition phase than after ignition, and flame retarding polymers so that ignition is prevented results in the polymers remaining in the smoke-producing phase short of ignition.

In some other cases, flame retarding the polymer succeeds in preventing oxidation which produces heat and flame, but the unoxidized material escapes into the gas phase in the form of soot and tar, forming smoke.

In the case of certain polymers such as polyvinyl chloride and polystyrene, the polymer is characterized by heavy smoke production

Table 5.2. Strengths of Chemical Bonds

Chemical Bond	Average Bond Energy kcal.	
C – H	98.2	
C – H	102	
C – Cl	78	
C – Br	65	
C – I	57	
H – H	103.2	
H – F	135	
H – Cl	102.1	
H – Br	86.7	
H – I	70.6	
F – F	37	
Cl – Cl	57.1	
Br – Br	45.4	
I – I	35.6	
C – C	77.7	ethane
	79.0	propane
	79.6	normal butane
	80.1	isobutane
	84.9	solid carbon
C = C	140.0	ethylene
C ≡ C	193.3	acetylene
C = C	123.8	benzene

over a wide range of conditions, unless compositions are specifically designed to reduce smoke.

Smoke retarding technology is not as extensively developed as flame retarding technology, and many additive compositions and formulations are proprietary.

Polyvinyl chloride tends to produce heavy smoke over a wide range of conditions and requires modification for smoke retardance. The approaches used are:

1. Use of fillers to dilute the polymer content, increase heat dissipation, and promote retention of material in teh char. Hydrated fillers such as alumina trihydrate dissipate more heat than nonhydrated fillers such as silica.

2. Use of iron compounds such as dicyclopentadienyl iron (ferrocene), ferric oxide, and ferric hydroxide.

3. Use of copper compounds such as cuprous cyanide and cuprous thiocyanate.

4. Use of vanadium compounds such as vanadium oxide and vanadium (III) acetylacetonate.

5. Use of molybdenum compounds such as molybdenum trioxide and zinc molybdates.

6. Use of magnesium and zinc compounds.

Polystyrene tends to produce heavy smoke and requires modification for smoke retardance. The approaches used are:

1. Use of fillers such as silica, alumina trihydrate, and hydrated clay.

2. Use of 8-hydroxyquinoline complexes of iron, manganese and chromium.

3. Use of phthalocyanine complexes of iron, copper, manganese, cobalt, and vanadium.

Acrylonitrile-butadiene-styrene (ABS) tends to produce heavy smoke and requires modification for smoke retardance. The approaches used are:

1. Use of fillers such as silica, alumina trihydrate, and hydrated clay.

2. Use of 8-hydroxyquinoline complexes of iron, manganese, and chromium.

3. Use of phthalocyanine complexes of iron, copper, manganese, cobalt, and vanadium.

4. Use of lead compounds such as tetraphenyl lead.

Polyesters tend to produce heavy smoke with some formulations and those formulations require modification for smoke retardance. The approaches used are:

1. Use of fumaric, maleic, adipic, succinic, or chlorendic acid in flame retardant formulations.

2. Use of phosphate compounds and barium sulfate.

3. Use of magnesium hydroxide in non-halogenated polyesters.

4. Use of molybdenum trioxide and ammonium molybdates.

Polyethylene and **polypropylene** tend to generate low levels of smoke, but some flame retardant formulations can produce increased smoke. The approaches used are:

1. Use of fillers such as alumina trihydrate.

2. Use of dicyclopentadienyl iron (ferrocene) and dicyclopentadienyl nickel (nickelocene).

3. Use of aluminum acetylacetonate and calcium carbonate.

Table 5.3. Chemical Extinguishing Agents

	Fire Extinguishing Effectiveness Weight Pct. Based on Ch_3Br	Explosion Suppression Effectiveness Weight Pct. Based on CCl_4
Carbon dioxide CO_2	124	
Carbon tetrachloride CCl_4 tetrachloromethane, Halon 104	68-75	100
Methyl bromide CH_3Br Halon 1001	100	192
Chlorobromomethane CH_2BrCl bromochloromethane, Halon 1011	80	180
Dibromodifluoromethane CBr_2F_2 Halon 1201, Freon 12B2	100-148	201
Bromotrifluoromethane $CBrF_3$ Halon 1301, Freon FE 1301, Freon 13B1	105-146	195
Bromochlorodifluoromethane $CBrClF_2$ Halon 1211, Freon 12B1	75	115
Dichlorodifluoromethane CCl_2F_2 Halon 122	68-104	98
Methylene bromide CH_2Br_2 Halon 1002		195
Bromodifluoromethane $CHBrF_2$		161
Ethyl bromide C_2H_5Br Halon 2001	50-96	
Dibromotetrafluoroethane $CBrF_2CBrF_2$ Halon 2402	74-107	139
Dibromochlorotrifluoroethane $CBrF_2CBrClF$ Halon 2312		59
Dibromodifluoroethane $CH_2BrCbrF_2$ Halon 2002		116

SECTION 5.5. SMOKE RETARDANCE MECHANISMS

Smoke retardance mechanisms have not been as extensively studied as have been flame retardance mechanisms, and only general mechanisms can be discussed here.

Certain compounds are believed to perform their smoke retardant function in the solid or condensed phase, by promoting char formation, diluting the polymer content, dissipating heat, and altering the chemical reactions in the condensed phase. This type of mechanism may be identified by an increase in the char residue.

Other compounds are believed to perform their smoke retardant function in the gas phase, by promoting oxidation of carbon or soot, and altering the chemical reactions in the gas phase.

SECTION 5.6. SMOLDER RETARDING VARIOUS POLYMERS

Smoldering combustion has been observed in cellulosic materials such as cotton fabrics and cotton batting, and cellulosic fiberboard and cellulose loose fiber insulation. It has been a significant problem in only two types of polymer materials: certain compositions of rigid phenolic foam, and certain compositions of flexible polyurethane foam.

Smoldering of flexible polyurethane foam occurs primarily as a result of smoldering upholstery fabrics. Many polyurethane compositions resist smoldering under these conditions; some exhibit smoldering when influenced by smoldering fabric; a few compositions propagate smoldering by themselves.

The approaches used to retard smoldering are:

1. Modification of the polymer composition, usually by modification of a proprietary formulation.

2. Modification of cell structure (porosity and cell size) to reduce susceptibility to smoldering.

SECTION 5.7. DESIGN OF THE APPLICATION

The use of plastic materials, to be safe from a flammability standpoint, must take into consideration both their behavior on exposure to fire and the probability and nature of possible fire exposure. The following general principles should be followed:

1. All highly flammable or easily ignited material should be removed from a potential source of heat or flame, and materials employed near this possible fire source should be tested to ensure that they have the necessary degree of flame retardance. The main example of this type of material, cellulose nitrate, should be replaced with other materials or adequately formulated with flame retardants.

2. Where possible, a material that is relatively easy to ignite should be protected with a coating or covering that is more difficult to ignite and that prevents access of oxygen to the easily ignited material.

3. Where possible, a material that exhibits relatively high surface flammability should be protected with a coating or covering of low surface

flammability. Because the thermal insulating qualities of many plastic foams prevent dissipation of heat and thus increase surface flame spread, plastic foams in general should not be applied without a protective coating or covering.

4. A material that exhibits relatively high surface flammability should not be applied continuously for great distances, and should not be applied in areas of high surface flame spread probability, such as ceilings. Regular firestops, either physical barriers or intervals of nonflammable surface, should be included in applications covering great distances.

5. A material that exhibits relatively high smoke production should not be exposed in large quantities or over large surface areas.

Ten rules presented for fire endurance rating by Harmathy are cited here because of their relevance to plastic materials used in structures:

1. The thermal fire endurance of a construction consisting of a number of parallel layers is greater than the sum of the thermal fire endurances characteristic of the individual layers when exposed separately to fire. A sandwich panel consisting of rigid polyuethane foam between polyvinyl chloride sheet facings, for example, provides more fire endurance than the individual components tests separately. The polyvinyl chloride sheet prevents the foam surface from being exposed to direct flame and limits access of oxygen, and the polyurethane foam provides support for the char from the polyvinyl chloride.

2. The fire endurance of a construction does not decrease with the addition of further layers. This rule is subject to various restrictions and should be carefully applied.

3. The fire endurance of constructions containing continuous air gaps or cavities is greater than the fire endurance of similar constructions of the same weight, but containing no air gaps or cavities. A polymer in cellular form exhibits greater fire endurance than the same weight of polymer in sheet form.

4. The farther an air gap or cavity is located from the exposed surface, the more beneficial is its effect on the fire endurance. Cellular plastics are more beneficial to fire endurance if they are not exposed to high heat, so that their thermal insulating qualities can be utilized without exposure to destructive temperatures.

5. The fire endurance of a construction cannot be increased by increasing the thickness of a completely enclosed air layer. This rule applies to air gaps of ½ inch or greater.

6. Layers of materials of low thermal conductivity are better utilized on that side of the construction on which fire is more likely to happen.

This rule is not applicable to materials such as plastics which undergo physical and chemical changes and evolve heat on burning.

7. The fire endurance of assymetrical constructions depends on the direction of heat flow. This rule is related to rules 4 and 6.

8. The presence of moisture, if it does not result in explosive spalling, increases the fire endurance. Water vapor has a beneficial effect in providing a noncombustible gas as well as requiring heat for vaporization. Oven-dried plastic materials tend to be more combustible than the same materials with equilibrium moisture. An extension of this principle is the use of encapsulated fluorocarbons to provide flame retardant action upon exposure to fire, and the incorporation of hydrated fillers.

9. Load-supporting elements, such as beams, girders, and joists, yielded higher fire endurances when subjected to fire endurance tests as parts of floor, roof, or ceiling assemblies than they would when tested separately.

10. The load-supporting elements of a floor, roof, or ceiling assembly can be replaced by such other load-supporting elements which, when tested separately, yielded fire endurances not less than that of the assembly.

SECTION 5.8. FIRE CONTROL AND EXTINGUISHMENT

In order to put out a fire it is necessary to stop the combustion reaction causing the fire. Fires are generally controlled and extinguished by one or more of the following methods:

1. cooling
2. separation or replacement of oxidizing agent
3. dilution or removal of fuel supply
4. chemical extinguishment

Extinguishment by cooling is the most widely used means of extinguishment, and the most effective in the case of ordinary combustible materials. In order to extinguish a fire by cooling, it is only necessary to absorb a portion of the total heat, since extinguishment occurs when the surface of the burning material is cooled to the point where it no longer releases enough vapors to maintain a combustible mixture. The efficiency of an extinguishing agent as a cooling medium depends on its specific and latent heats, and in this regard water is particularly effective since its specific and latent heats are higher than those of most extinguishing agents.

Extinguishment by separation of the oxidizing agent is accomplished by blanketing or smothering a fire. Carbon dioxide, foam, and vaporizing liquids form a blanket which prevents oxygen from reaching the fire. If the blanket is maintained long enough to cool the material below its self-ignition temperature and there are no ignition sources, the fire will stay out. It is, however, difficult to maintain the blanket long enough to extinguish all smoldering sources of ignition, and extinguishment by separation cannot be accomplished with materials containing their own oxygen supply, such as cellulose nitrate.

Chemical extinguishment is based on the observed effectiveness of certain halogenated hydrocarbons and inorganic salts as extinguishing agents. This effectiveness is greater than can be accounted for by cooling, smothering, or blowing out mechanisms. The commercially used extinguishing agents are strikingly similar in chemical nature to the flame retardants discussed in an earlier section.

CHAPTER 6

Market Acceptance Criteria

SECTION 6.1. MARKETS FOR THE PLASTICS INDUSTRY

Plastic materials have gained general acceptance because they contribute to the fulfillment of human needs and human desires. The markets for these materials can be divided accordingly, into four broad areas.

1. Building construction: where man lives and works.
2. Building contents: what man uses for his comfort and convenience.
3. Transportation: how man travels and moves his belongings.
4. Industrial goods: what man uses in his work.

The factors which determine acceptance of any material into any market are many; these include economics, politics, salesmanship, and personal appeal. The principal factor, however, is performance. The material must meet the requirements of the application. If a material fails to meet performance requirements, blaming lack of acceptance on other factors may be soothing but definitely is irrelevant. The most favorable purchasing agent, the most receptive building code, the lowest raw material costs can not make an unsatisfactory material work.

The fact that plastic materials are so thoroughly entrenched in many applications indicates that plastic materials do meet the requirements of those applications. In the majority of cases, the plastic material was designed for the application, and the essential requirements of the application were not altered for the plastic material. This situation holds true for flame retardance. Each market, and each application in that market, has a requirement for a specific level of flame retardance, depending on experience and judgement.

Because of population growth and population shifts, conditions in all these markets, particularly building construction, are changing rapidly and generating presures for changes in many requirements affecting plastic materials. While these pressures will doubtless tend to make acceptance requirements more reasonable, the requirement of adequate safety to human life and property will never be abolished, and increasing government interest is making certain that safety will not be neglected.

SECTION 6.2. TYPES OF ACCEPTANCE CRITERIA

For a plastic material to win acceptance in any market, it must meet three types of requirements: customer requirements, government requirements, and insurance requirements. These are not necessarily independent, and in some cases one type of requirement dominates and influences acceptance of the material.

Customer requirements are based on the basic tenets of our free-enterprise system. The customer is supposedly free to set his own requirements and to purchase materials from any supplier who meets those requirements. Since the customer is not always the ultimate consumer, customer requirements sometimes include not only his own requirements but also the requirements of his own customers. Where use of the material in a particular application is subject to government and insurance regulations, these additional requirements are often incorporated in the customer's specifications. Where the customer is part of a trade association, such as the Mobile Home Manufacturers Association, the quality standards agreed upon by members of the association will often be incorporated into the customer's requirements.

Government requirements are based on the government's function of promoting the safety and welfare of its citizens. Government requirements have increased greatly in importance for the following reasons: first, the rise of the government as a large customer itself, due to its massive expenditures in the military and aerospace fields, and the tendency toward centralization of standard-setting for government purchases of items in quantity, such as motor vehicles and office supplies; second, the widespread use of influence on specific sectors of government to promote special interests, such as the efforts of plumbers' unions and cast iron trade associations to block acceptance of plastic pipe and fittings in local building codes; third, the widespread failure of the various industries to exhibit adequate concern for the public welfare, by placing on the market materials such as highly flammable fabrics, and interior finishes with high surface flame spread characteristics.

Government requirements fall into four groups: federal, military, state, and local.

Insurance requirements arise from the general practice of purchasing insurance for protection against catastrophic losses, and the consequent desire of the insurance companies to protect themselves against excessive risks, or risks not justified by the premiums charged. Since insurance premiums are part of the cost of doing business, insurance rates have a direct influence on profitability.

SECTION 6.3. THE BUILDING CONSTRUCTION MARKET

The building construction market can be divided into four areas: residential (homes, apartments, mobile homes), commercial (business mercantile, warehouses), industrial (plants, pipe lines), and transportation (highways, harbors, airports). The nature of this market makes it difficult to divide plastics consumption according to these areas, since building materials suppliers serve two or more areas. The present consumption of plastics, however, is sizable, and the potential is large. Total consumption of plastics in building construction was over 3 million metric tons in 1980 (Table 6.1).

There are about 18,000 building materials supply companies in the United States, and they in effect control the selection of materials for this highly fragmented industry. The architects and engineers involved exercise some choice in specifying materials, but the overriding influence in customer requirements is the building code involved, since the building code restricts the architect's freedom in design.

Government requirements on the federal level come from a variety of departments: the Department of Housing and Urban Development (HUD), particularly the Federal Housing Administration (FHA); the Department of Health, Education, and Welfare (HEW); the Department of Transportation (DOT); the General Services Administration (GSA).

The military specifications pertinent to the construction market have been assigned to the following groups in the Federal Supply Classification:

Table 6.1. Uses of Plastics in Building Construction

	Consumption 1000 metric tons	
	1979	1980
Pipe, fittings, conduit		
Polyvinyl chloride (PVC)	1,114	938
High density polyethylene	244	209
Acrylonitrile-butadiene-styrene (ABS)	150	110
Reinforced polyester	107	100
Polypropylene	11	11
Low density polyethylene	15	9
Epoxy coatings	4	4
Polystyrene	5	3
Resin-bonded woods		
Urea and melamine	502	435
Phenolic	232	210
Insulation		
Phenolic binder	131	128
Rigid urethane foam	133	120
Polystyrene foam	93	80
Flooring		
Polyvinyl chloride (PVC)	164	114
Urethane foam carpet underlay	60	50
Epoxy	9	8
Panels and siding		
Polyvinyl chloride (PVC)	106	92
Reinforced polyester	70	65
Acrylic	6	5
Butyrate	2	2
Vapor barriers		
Low density polyethylene	78	70
Polyvinyl chloride (including pool liners)	14	14
Plumbing		
Reinforced polyester	63	60
Acrylic	11	10
Polyacetal	8	8
Thermoplastic polyester	3	3
Polystyrene	2	2
Glazing and skylights		
Polycarbonate	32	31
Acrylic	33	30
Reinforced polyester	18	15

(Continued)

Table 6.1. Uses of Plastics in Building Construction (Continued)

	Consumption 1000 metric tons	
	1979	1980
Profile extrusions		
Polyvinyl chloride (including foam)	53	53
Polyethylene	3	2
Decorative laminates		
Phenolic	24	20
Urea and melamine	13	12
Wall coverings		
Polyvinyl chloride (PVC)	25	22
Polystyrene	6	4
Lighting fixtures		
Acrylic	10	9
Polystyrene	12	7
Polyvinyl chloride (PVC)	8	7
Polycarbonate	5	4
Cellulosics	2	2
Total	3,581	3,078

(Concluded)

Group 45	Plumbing, Heating, and Sanitation Equipment
Group 47	Pipe, Tubing, Hose, and Fittings
54	Prefabricated Structures and Scaffolding
55	Lumber, Millwork, Plywood, and Veneer
56	Construction and Building Materials
61	Electric Wire, and Power and Distribution Equipment
93	Nonmetallic Fabricated Materials
	9330 Plastics Fabricated Materials
95	Metal Bars, Sheets, and Shapes

Government requirements are to a large degree embodied in building codes enforced by local building officials. Since each building code body is independent, a material of construction must obtain approval within every locality within which it is to be used. To provide guidance for the smaller cities and towns which cannot support extensive research and testing facilities, and to encourage uniformity, several model building codes have been prepared by various organizations. The most important ones are:

National Building Code
  National Board of Fire Underwriters, New York
Uniform Building Code
  International Conference of Building Officials, Los Angeles
Southern Standard Building Code
  Southern Building Code Congress, Birmingham, Alabama
Basic Building Code
Building Officials Conference of America, Chicago
Modern Standard Building Code
  Midwest Conference of Building Officials, Chicago
National Building Code – Canada
  National Research Council, Ottawa, Canada

Perhaps the three most widely accepted model building codes in the United States are the Basic Building Code (BOCA) for the northeast and midwest, the Uniform Building Code (ICBO) for the west, and the Southern Standard Building Code for the south. They have been adopted to a substantial extent but not in their entirety, since each building code group is independent and influenced by the most persuasive special-interest groups in its locality.

There are two ways of classifying buildings: according to construction, (Table 6.2) and according to occupancy (Table 6.3). Depending on the code, there may be four to six construction classifications, and the most highly-rated fire-resistance classifications involve the use of steel, iron, concrete, and masonry. These materials are noncombustible and backed by years of experience; their continued use is promoted by their respective trade associations, and guarded by labor unions: ironworkers, bricklayers, and cement workers. The least fire-resistant classifications involve the use of wood, a material which is combustible but entrenched by centuries of use; its continued use is promoted by its trade associations and guarded by carpenters' unions.

Despite the generally conservative appearance of the building codes, they are more receptive to new materials and techniques than is immediately apparent, because they are to a substantial extent performance-oriented. The requirements are largely presented on a performance basis, and while these requirements may seem extremely demanding to the plastics industry, two features of these requirements must be pointed out: they are more or less justified by concern for public safety, and they are now being met by commercially available materials.

Any material, plastic or otherwise, can be classified as noncombustible if it meets one of the following requirements:

1. It does not liberate flammable gases when heated to 1382°F. according to ASTM E 136; or

2. It has a structural base on noncombustible material as defined in (1) and a surfacing not over 1/8 inch thick with a flame spread rating not higher than 50 according to ASTM E 84; or

3. It has a surface flame spread rating not higher than 25 according to ASTM E 84, and is of such composition that surfaces that would be exposed by cutting through the material in any way would not have a flame spread rating higher than 25.

Table 6.2. Construction Classifications

Basic Building Code	
Building Officials Conference of American (BOCA)	
Type 1	Fireproof Construction
Type 2	Noncombustible Construction
Type 3	Exterior Masonry Wall Construction
Type 4	Frame Construction
Uniform Building Code (UBC)	
International Conference of Building Officials (ICBO)	
Type I	Noncombustible: steel, iron, concrete, masonry
Type II	Noncombustible: steel, iron, concrete, masonry
Type III	Combustible: one-hour or heavy timber
Type IV	Noncombustible: noncombustible materials
Type V	Combustible: any materials allowed by this code
Southern Standard Building Code	
Southern Building Code Congress (SBCC)	
Type I	Fireproof
Type II	Fire-Resistive
Type III	Heavy Timber
Type IV	Noncombustible Frame
Type V	Ordinary
Type VI	Wood Frame

In the Southern Standard Building Code, the maximum allowable flame spread rating for the surfacing in definition (2) is 25.

Fire resistance ratings are generally defined according to ASTM E 119. Since no commercially available plastic materials can provide significant fire resistance as a structural component, plastics will find their greatest use in applications which do not require fire resistance ratings: non-

Table 6.3. Occupancy Classifications

Basic Building Code		
Building Officials Conference of America (BOCA)		
A		high hazard, involving highly combustible or explosive products or materials, or highly corrosive or poisonous materials
B		storage, including warehouses and freight depots
	B-1	moderate hazard, involving contents which burn with moderate rapidity, but are neither poisonous nor explosive
	B-2	low hazard, involving noncombustible materials or materials which do not burn rapidly
C		mercantile, including retail stores, shops, and markets
D		industrial, including factories, assembly plants, and laboratories
E		business, including offices, banks, testing laboratories, radio stations, and motor fuel service stations
F		assembly
	F-1	theaters
	F-2	night clubs, dance halls
	F-3	museums, libraries, restaurants, recreation centers, passenger terminals
	F-4	churches, schools, colleges
	F-5	coliseums, stadiums, drive-in theatres
H		institutional
	H-1	restrained occupants, including prisons, jails, reformatories, and insane asylums
	H-2	incapacitated occupants, including hospitals, clinics, homes for the aged and infirm; and emergency services, including police stations and fire houses
L		residential
	L-1	accommodating more than 20 individuals, including hotels, boarding houses, and dormitories
	L-2	accommodating less than 20 individuals; and all multiple-family dwellings, including apartments
	L-3	one- or two-family dwelling units, including mobile homes
M		miscellaneous
Uniform Building Code (UBC)		
International Conference of Building Officials (ICBO)		
A		assembly building with a stage, and an occupant load of 1000 or more

(Continued)

Table 6.3. Occupancy Classifications (Continued)

B	B-1	assembly building with a stage, and an occupant load of less than 1000
	B-2	assembly building without a stage, and an occupant load of 300 or more
	B-3	assembly building without a stage, and an occupant load of less than 300
	B-4	stadiums, reviewing stands, and amusement park structures
C		school or day-care purposes more than 8 hours per week, involving assemblage for instruction, education or recreation, and not class in A, B-1, or B-2
D	D-1	mental hospitals, jails, prisons, reformatories
	D-2	hospitals, sanitariums, nurseries, nursing homes for nonambulatory patients
	D-3	nursing homes for ambulatory patients
E	E-1	storage and handling of hazardous and highly flammable or explosive materials other than flammable liquids
	E-2	storage and handling of flammable liquids
	E-3	buildings where loose combustible fiber or dust is manufactured, processed, or generated; storage of highly combustible materials
	E-4	repair garages
	E-5	aircraft repair hangers
F	F-1	gasoline service stations
	F-2	wholesale and retail stores, office buildings, drinking and dining establishments having an occupant load of less than 100, police and fire stations, factories and workshops not handling highly flammable or combustible materials
	F-3	aircraft hangars where no repair work is done
G		ice plants, power plants, pumping plants, cold storage, factories and workshops using noncombustible and nonexplosive materials, storage of noncombustible and nonexplosive materials
H		hotels, apartment houses, convents, monasteries
I		dwellings and lodging houses
J	J-1	private garages, sheds, and agricultural buildings not over 1000 square feet in area
	J-2	fences over 6 feet high, tanks, and towers
Southern Standard Building Code		
Southern Building Code Congress		
A		residential, including dwellings, multiple dwellings, hotels, dormitories, lodging houses, convents, and monasteries

(Continued)

Table 6.3. Occupancy Classifications (Continued)

B	business
B-1	office buildings, greenhouses, banks, service stations
B-2	stores, shops, markets, restaurants
C	schools, including colleges, universities, academies
D	institutional
D-1	insane asylums, reformatories, jails, prisons
D-2	hospitals, sanitariums, orphanages, old peoples homes
E	assembly, including passenger depots, libraries, stadiums, theaters, churches, museums, auditoriums
E-1	large assembly, having a working stage and a capacity of 700 or more people, or having a non-working stage and a capacity of 1000 or more people
E-2	small assembly, having a capacity of 75 or more persons but less than designated for E-1
F	storage, excluding highly combustible, flammable or explosive products
G	industrial, excluding highly combustible, flammable, or explosive products or materials
H	special hazardous, involving highly combustible, flammable, or explosive products or materials

(Concluded)

structural uses in noncombustible construction, and both structural and non-structural uses in combustible construction; in short, direct competition with wood and glass.

The occupancy classifications of buildings are a measure of relative hazard to life, since they indicate both the number and mobility of the persons in the building. The formal classifications vary somewhat (Table 6.3), but the various building codes are based on similar density of occupancy for similar occupancy types (Table 6.4).

Interior finish materials, a substantial application area for plastic materials, are presently classified for fire hazard on the basis of flame spread ratings by ASTM E 84. Classifications vary with the building code (Table 6.5), but requirements tend to be similar for similar occupancies (Table 6.6). The most demanding requirements are applied for exitways, and for occupants under restraint in institutions, and the Basic Building Code and the Uniform Building Code are more stringent in this regard than the Southern Standard Building Code.

The emphasis on flame spread rating for exitways in particular and all occupancies in general is the result of several tragic fires involving great

Table 6.4. Available Area per Occupant (square feet per occupant)

Code Organization	**BBC BOCA**	**UBC ICBO**	**SSBC SBCC**	**BEC NFPA**
Assembly, fixed seats	6	7	6	
Assembly, without fixed seats	15	15	15	15
Educational	40	20	40	40
Institutional	150	80	125	75
Residential, dwellings		300	125	
Residential, hotels and apartments	125	200	125	125
Mercantile, basement	30	20	30	30
Mercantile, ground floor	30	30	30	30
Mercantile, upper floors	60	50	60	60
Business, office	100	100	100	100
Industrial			100	100
Storage	300	300	300	300
Aircraft hangars		500		

Table 6.5. Interior Finish Classifications Based on ASTM E 84 Flame Spread

Basic Building Code		
Building Officials Conference of America (BOCA)		
	I	0 to 25
	II	26 to 75
	III	76 to 200
	IV	over 200
Uniform Building Code		
International Conference of Building Officials (ICBO)		
	I	0 to 25
	II	26 to 75
	III	76 to 225
Building Exits Code		
National Fire Protection Association (NFPA)		
	A	0 to 25
	B	25 to 75
	C	75 to 200
	D	200 to 500
	E	over 500

Table 6.6. Interior Finish Requirements: Maximum ASTM E 84 Flame Spread

Basic Building Code	
Building Officials Conference of American (BOCA)	
required vertical exitways and passageways	
high hazard, storage, mercantile, industrial, business, assembly, institutional, residential except one- and two-family dwellings	25
corridors providing access to exitways	
theaters, nightclubs, institutional,	
high hazard, storage, mercantile, industrial business, terminals, restaurants, churches, schools, residential except one- two-family dwellings	75
one- and two-family dwellings	200
rooms or enclosed spaces	
institutional	25
theaters, nightclubs	75
high hazard, storage, mercantile, industrial, business, terminals, restaurants, churches, schools, residential except one- and two-family dwellings	200
one- and two-family dwellings	500
Uniform Building Code	
International Conference of Building Officials (ICBO)	
enclosed vertical stairways	
all occupancies except dwellings, lodging houses, private garages, and tanks and towers	25
no requirements for dwellings and lodging houses	
other exit ways	
all occupancies except dwellings, lodging houses, private garages, tanks and towers	75
no requirements for dwellings and lodging houses	
rooms or areas	
institutional, forcible restraint	25
institutional, other than forcible restraint	75
all other occupancies except dwellings, lodging houses, private garages, tanks and towers	225
no requirements for dwellings and lodging houses	
Southern Standard Building Code	
Southern Building Code Congress	
vertical enclosures in buildings more than three stores high	75
required horizontal enclosures	200
rooms or spaces in institutional buildings where occupants are restrained	75
other rooms or spaces in institutional buildings	200
assembly, school, business, mercantile, apartment, and hotel occupancies	200

loss of life. One factor contributing to the death of 492 persons in the Cocoanut Grove night club fire in Boston in 1942 was rapid flame spread; the ceiling finish material in the room had a flame spread rating of 2,500. In general, a plastic material with a flame spread rating over 200 has a limited future in the construction market. A rating under 200 makes a portion of the market accessible, and the ratings of 130 to 190 given for some plastic panels indicate that the market justifies listing of these particular products. A rating under 75 makes available another portion of the market, and a rating under 25 gains the highly desirable noncombustible classification.

For applications other than structural members and interior finish materials, the "approved plastics" classification is pertinent. The Uniform Building Code defines an approved plastic material as one which has a flame spread rating of 225 or less and a smoke density not greater than that obtained from the burning of untreated wood under similar conditions. The Southern Standard Building Code lists as a basis for approval a burning rate not exceeding 2.5 inches per minute in the ASTM D 635 test, or a burning time not less than 2 minutes for complete consumption in the ASTM D 568 test.

Government requirements on the state level are generally within the jurisdiction of state fire marshals, who usually maintain close contact with local building code officials and insurance groups. In the larger states such as New York and California, school officials often have set requirements for the approval of materials to be used in schools.

Insurance requirements are set by individual underwriters, and the insurance companies generally look to their technical experts for guidance in the field of fire hazard. Underwriters Laboratories (UL) serves as the technical organization of the Factory Insurance Association (FIA), while the Factory Mutual Engineering Corporation is the technical organization of the Associated Factory Mutual Insurance Companies.

SECTION 6.4. THE BUILDING CONTENTS MARKET

The building contents market, or consumer goods market, can be divided into six general areas: wearing apparel, interior furnishings, housewares, appliances, packaging, and miscellaneous.

Polymers in the physical shape of fibers are produced in substantial volumes for this market, but production and consumption figures for the uses of various polymers are difficult to obtain in the highly competitive fibers and fabrics industry, partly because of the substantial volumes of imports and exports.

In the wearing apparel area, clothing applications in 1980 consumed significant quantities of plastic materials: 38,000 metric tons of polyvinyl chloride for footwear, 5,000 metric tons of polyvinyl chloride for outerwear, and 2,000 metric tons of polyvinyl chloride for baby pants.

In the interior furnishings area, furniture applications in 1980 consumed 409,000 metric tons of plastics, including 220,000 metric tons of flexible urethane foam and 77,000 metric tons of polyvinyl chloride (Table 6.7). Bedding consumed 80,000 metric tons of flexible urethane foam. Houseware applications consumed 491,000 metric tons of plastics, including 93,000 metric tons of polystyrene, 77,000 metric tons of polypropylene, 175,000 metric tons of low-density polyethylene, and 66,000 metric tons of high-density polyethylene (Table 6.8). Carpet underlay applications consumed 50,000 metric tons of flexible urethane foam.

Table 6.7. Uses of Plastics in Furniture

	Consumption 1000 metric tons	
	1979	**1980**
Polyurethane flexible foam	230	220
Polyvinyl chloride (PVC)	98	77
Polystyrene	63	50
Phenolic for plywood	21	19
Polypropylene	19	16
Phenolic for decorative laminates	11	10
Polyurethane rigid foam	12	8
Polyethylene	9	8
Melamine	8	7
Other polyurethane	5	4
Acrylonitrile-butadiene-styrene (ABS)	6	4
Polyester	2	2
Other	4	4
Total	488	409

Materials used in upholstered furniture sold in the State of California must comply with the requirements of Technical Information Bulletins 116 and 117 issued by the California Bureau of Home Furnishings. Requirements for matresses in high risk occupancies such as correctional institutions are contained in Technical Information Bulletin 121.

Table 6.8. Uses of Plastics in Housewares

	Consumption 1000 metric tons	
	1979	**1980**
Low density polyethylene	216	175
Polystyrene	114	93
Polypropylene	77	77
High density polyethylene	73	66
Polyvinyl chloride (PVC)	23	21
Phenolic	20	18
Melamine	18	13
Styrene acrylonitrile	9	8
Other	22	20
Total	572	491

The area of appliances, including major appliances such as refrigerators and ranges, radio and television, and minor appliances, consumed 330,000 metric tons of plastics in 1980, including 51,000 metric tons of polystyrene, 60,000 metric tons of acrylonitrile-butadiene-styrene, 39,000 metric tons of polypropylene, 17,000 metric tons of phenolic, and 29,000 metric tons of rigid urethane foam (Table 6.9).

The miscellaneous group of applications includes toys, records, garden hose, and luggage. Toys and novelties alone consumed 243,000 metric tons of plastics in 1980, including 65,000 metric tons of polystyrene, and 93,000 metric tons of polyethylene, both low-density and high-density (Table 6.10). Records consumed 52,000 metric tons of polyvinyl chloride, and garden hose consumed 16,000 metric tons of polyvinyl chloride. Luggage and cases consumed 10,000 metric tons of acrylonitrile-butadiene-styrene.

The companies which are customers for plastic materials have a variety of requirements, and it would be a monumental task to discuss all flammability requirements for various applications. In many cases, the principal flammability requirements are those resulting from government action.

Government requirements on the federal level come from a variety of departments and agencies: the Department of Commerce (DOC), particularly the National Bureau of Standards (NBS); the Department of Health, Education, and Welfare (HEW), particularly the Public Health Service (PHS) and the Food and Drug Administration (FDA); the General

Table 6.9. Uses of Plastics in Appliances

	Consumption 1000 metric tons	
	1979	**1980**
Acrylonitrile-butadiene-styrene (ABS)	73	60
Polystyrene	65	51
Reinforced polyester	52	40
Polypropylene	45	39
Polyvinyl chloride (PVC)	39	32
Rigid polyurethane foam	33	29
Phenolic	20	17
Modified polyphenylene oxide	15	14
Polycarbonate	15	13
Styrene acrylonitrile	8	7
Nylon	8	6
Acrylic	4	3
Epoxy	4	3
Polyacetal	4	3
Polyethylene	4	3
Thermoplastic polyester	2	2
Cellulosics	1	1
Other	9	7
Total	401	330

Table 6.10 Uses of Plastics in Toys and Novelties

	Consumption 1000 metric tons	
	1979	**1980**
Polystyrene	73	65
Low density polyethylene	62	56
High density polyethylene	41	37
Polypropylene	32	34
Polyvinyl chloride (PVC)	18	15
Acrylonitrile-butadiene-styrene (ABS)	13	12
Cellulosics	2	1
Other	22	23
Total	263	243

Services Administration (GSA), particularly the Federal Supply Service (FSS); the Consumer Product Safety Commission (CPSC); the Federal Trade Commission (FTC).

The military specifications pertinent to consumer goods have been assigned to the following groups in the Federal Supply Classification:

Group 41	Refrigeration and Air Conditioning Equipment
45	Plumbing, Heating, and Sanitation Equipment
62	Lighting Fixtures and Lamps
71	Furniture
72	Household and Commercial Furnishings and Appliances
73	Food Preparation and Serving Equipment
77	Musical Instruments, Phonographs, and Radios
78	Recreational and Athletic Equipment
83	Textiles, Leather, Furs, Apparel, and Shoes
84	Clothing, Individual Equipment, and Insignia
85	Toiletries

Insurance requirements are pertinent only to identifiable items which can be labeled for approval, such as home appliances. The Underwriters Laboratories (UL) label of approval is a prominent symbol in this regard. Insurance requirements on other products generally involve storage hazards in warehouses.

SECTION 6.5. THE TRANSPORTATION MARKET

The transportation market can be divided into five general areas: automotive, principally passenger cars; highway transportation, including trailer and tank trucks; railway transportation, including passenger coaches, freight cars, and tank cars; aircraft, including aerospace applications; and marine transportation, including boats, barges, and ocean-going vessels.

The automotive area is the largest consumer of plastic materials in the transportation market, accounting for 708,000 metric tons of plastics in 1980 model passenger cars. This quantity included 160,000 metric tons of polyurethanes, 150 metric tons of reinforced polyester, 110,000 metric tons of polyvinyl chloride, and 140,000 metric tons of polypropylene and copolymers (Table 6.11).

Table 6.11. Uses of Plastics in Transportation

Model Year	Consumption 1000 metric tons 1980	1981
Passenger cars		
Polyurethane	160	175
Reinforced polyester	150	170
Polypropylene	140	150
Polyvinyl chloride	110	120
ABS	55	57
Nylon	24	29
Acrylic	18	19
Phenolic	15	16
Other	36	40
Total	708	776
Trucks, buses, rail transport, aircraft		
Polyurethane	28	30
Reinforced polyester	20	22
Polypropylene	14	15
Polyvinyl chloride	10	11
Acrylic	9	10
ABS	6	7
Nylon	4	5
Phenolic	2	3
Other	4	5
Total	97	108

The automobile industry in recent years has produced from 6 to 10 million passenger cars annually. The production of trucks and buses involves about 2 million vehicles annually. Over one hundred thousand truck trailers are produced each year.

The production of freight-train and passenger-train cars does not follow as steady an annual pattern as the production of motor vehicles and highway trailers, since each railroad company has its own schedule for replacements and additions to its rolling stock.

Because passenger aircraft have no provision for egress in the event of an in-flight fire, the fire safety aspects of polymeric materials used in aircraft have been a prominent flammability problem in the transportation market.

Government requirements in the transportation market are to a substantial extent determined by the Department of Transportation. The Federal Highway Administration has the responsibility for motor vehicles and highway transportation, the Federal Railroad Administration for railway transportation, the Federal Aviation Administration for air transportation and aircraft, and the Coast Guard for ships and boats.

The pertinent rules and regulations issued by the Coast Guard are included in the following publications:

CG-256	Rules and Regulations for Passenger Vessels
	Part 72. Construction and Arrangement
	Part 75. Lifesaving Equipment
CG-323	Rules and Regulations for Small Passenger Vessels
CG-257	Rules and Regulations for Cargo and Miscellaneous Vessels
	Part 92. Construction and Arrangement
	Part 94. Lifesaving Equipment
CG-123	Rules and Regulations for Tank Vessels
CG-290	Pleasure Craft

The military specifications pertinent to the transportation market have been assigned to the following groups in the Federal Supply Classification:

Group 15	Aircraft and Airframe Structural Components
16	Aircraft Components and Accessories
18	Space Vehicles
19	Ships, Small Craft, Pontoons, and Floating Docks
20	Ship and Marine Equipment
22	Railway Equipment
23	Motor Vehicles, Trailers, and Cycles
24	Tractors
25	Vehicular Equipment Components

SECTION 6.6. THE INDUSTRIAL GOODS MARKET

The industrial goods market can be divided into four general areas: machinery and machine parts, electrical and electronic, coatings, and packaging.

The electrical and electronics area consumes large volumes of plastics for wire and cable coatings, electrical controls and switchgear, and communications equipment. Consumption in 1980 totalled 736,000 metric tons, including 211,000 metric tons of polyolefins and 177,000 metric tons of polyvinyl chloride for wire and cable, 63,000 metric tons of phenolic, and 120,000 metric tons of styrenics (Table 6.12).

Table 6.12. Uses of Plastics in Electrical and Electronics Markets

	Consumption 1000 metric tons	
	1979	**1980**
Polyvinyl chloride for wire and cable	195	177
Low density polyethylene for wire and cable	193	161
Polystyrene	123	105
Reinforced polyester	80	68
Phenolic	69	63
High density polyethylene for wire and cable	57	50
Polycarbonate	22	21
Acrylonitrile-butadiene-styrene (ABS)	18	15
Urea	18	14
Nylon (including wire and cable)	15	13
Epoxy for electrical laminates	11	11
Modified polyphenylene oxide	12	10
Thermoplastic polyester	9	9
Polypropylene	5	5
Polyacetal	2	2
Styrene acrylonitrile	2	2
Cellulosics	1	1
Other	9	9
Total	841	736

Industrial coatings and packaging applications are not as thoroughly documented because the consumption figures are spread over two or more markets.

The complexity of this market makes a listing of all flammability requirements impossible in this publication. In many cases, there are no flammability requirements other than those based on government or insurance requirements. Materials that are difficult to ignite appear to be generally acceptable.

Government requirements on the federal level are largely based on the government's position as a major purchaser, and generally come from the General Services Administration (GSA), particulary the Federal Supply Service. Various government offices, however, indirectly influence requirements.

The Mine Safety and Health Administration requires electrical cable and conveyor belting and hose to pass flame resistance tests described in Sections 18.64 and 18.65 of Title 30, Part 18.

The military specification pertinent to the industrial goods market have been assigned to the following groups in the Federal Supply Classification:

Group 28	Engines, Turbines, and Components
29	Engine Accessories
30	Mechanical Power Transmission Equipment
31	Bearings
32	Woodworking Machinery and Equipment
34	Metalworking Machinery
35	Service and Trade Equipment
36	Special Industry Machinery
37	Agricultural Machinery and Equipment
38	Construction, Mining, Excavating, and Highway Maintenance Equipment
39	Materials Handling Equipment
40	Rope, Cable, Chain, and Fittings
41	Refrigeration and Air Conditioning Equipment
43	Pumps and Compressors
44	Furnace, Steam Plant, and Drying Equipment; and Nuclear Reactors
46	Water Purification and Sewage Treatment Equipment

Group	47	Pipe, Tubing, Hose, and Fittings
	48	Valves
	49	Maintenance and Repair Shop Equipment
	58	Communication, Detection, and Coherent Radiation Equipment
	59	Electrical and Electronic Equipment Components
	61	Electric Wire, and Power and Distribution Equipment
	66	Instruments and Laboratory Equipment
	67	Photographic Equipment
	74	Office Machines, Visible Record Equipment, and Data Processing Equipment
	81	Containers, Packaging, and Packing Supplies

Insurance requirements on particular products in this market have been prepared by Underwriters Laboratories (UL) and Factory Mutual (FM).

CHAPTER 7

Flammability and Product Liability

Product liability litigation has become one of the major economic factors in product safety. The increasing number of claims, the increasing size of awards made by juries, and the increasing cost of litigation, all together have multiplied the cost of this type of litigation to manufacturers and their insurance companies. The insurance companies have increased the premiums for liability insurance accordingly, and the cost of increased premiums becomes part of the price of the finished product when sold in commerce.

There are at least two opposing philosophies regarding product liability litigation. Opponents of such litigation believe that the substantial resources expended in ligitation could be better spent in making the product safer. Proponents of such litigation believe that only the economic penalty of prohibitive jury awards is effective in forcing manufacturers to improve the safety of their products.

SECTION 7.1. HISTORY OF PRODUCT LIABILITY LITIGATION

Product liability litigation has developed gradually over the years from a relatively simple matter into a sophisticated technology. Under the rule of privity, a seller was liable for injury by his product only to the buyer to whom he contracted to sell the product, and this rule was involved in the case of Winterbottom v Wright in 1842. In the case of Brown v Kendall in 1850, it was held that the plaintiff had to prove negligence by the defendant in order to impose liability for accidental injury.

Two exceptions to the rule of privity began to decrease the protection which the privity rule afforded manufacturers. In the case of Thomas v Winchester in 1852, products considered imminently dangerous because of the manufacturer's negligence made the seller liable for injury regardless of whether there was a contract. In the case of Huset v J. I. Case in 1903, a manufacturer who delivered a product without giving notice of a known dangerous condition was held to be liable to anyone injured in using that product. For most practical purposes, the rule of privity was essentially destroyed in 1916 in the case of MacPherson v Buick Motor Company, when it was held that the manufacturer was liable if there was negligence in the assembly of the product, and was liable to persons, other than the purchaser, who used the product.

An express warranty theoretically limits the manufacturer's liability to replacement of the defective part or assembly. In the case of Henningen v Bloomfield Motors in 1960, it was held that each product, by being put on the market for sale, bears an implied warranty that is reasonably safe for use. This 1960 decision is considered by some to be the modern beginning of the principle of strict liability, which was established in 1963 by the California Supreme Court in the case of Greenman v Yuba Power Products. Under this doctrine, a manufacturer is strictly liable when the product proves to have a defect that causes injury to a human being, and the plaintiff does not need to prove negligence.

In the case of Noel v United Aircraft in 1964, it was ruled that the supplier of a product can be found liable by failing to make a newly developed safety device available to owners of the product, or by failing to warn them of a dangerous condition discovered after the product was delivered.

In the case of Bart v B. F. Goodrich Tire Company, it was ruled that contributory negligence of the plaintiff or others is no defense in a strict liability action. The degree of the plaintiff's negligence, generally less than 50 percent, is used to pro-rate the liabilities.

In the case of Elmore v American Motors in 1969, it was ruled that even a bystander could sue if injured, and could sue not just the operator of a car but even the car manufacturer. When the third-party suit is used against third-parties capable of paying off large claims, either in judgements or settlements, the philosophy is sometimes called the deep-pocket principle.

In the case of Thomas v General Motors in 1970, the manufacturer was held liable for any and all foreseeable intended uses, misuses and abuses of the product and even abnormal uses which were foreseeable and could have been safeguarded against.

In the case of Cronin v J. B. E. Olson Corporation in 1972, it was ruled that a product being defective and resulting in an injury is enough to impose strict liability on the manufacturer, and that the plaintiff's awareness of the defect is no defense.

In the case of Balido v Improved Machinery in 1973, it was ruled that warnings and directions do not absolve the manufacturer of liability if there is a defect in design or manufacture. In the case of Glass v Ford Motor Company, also in 1973, it was ruled that the burden of determining whether the defective part was due to a manufacturing or design defect is no longer imposed on the plaintiff.

Two relatively new theories of liability have been advanced. Under the theory of enterprise liability or industry-wide liabity, suggested in the 1972 case of Hall v E. I. du Pont de Nemours and Company, if the defendants, essentially all the companies in an industry and its trade association, had adhered to an industry-wide standard concerning the safety features, manufacture, and design of the product, if the plaintiffs could trace the product to one of the defendants, then the burden of proof as to causation would shift to all the defendants, and each industry member would have contributed to the plaintiff's injury. Under the theory of marekt share liability, adopted in a ruling of the California Supreme Court in 1980 in the case of Sindell v Abbott Laboratories, each defendant is held liable for a percentage of the judgement, as determined by that defendant's market share, unless a particular defendant can demonstrate that it could not have made the product that injured the plaintiff.

The contingency fee basis used in many product liability cases makes many more claims possible than would be the case if injured parties had to rely only on their own resources. In a contingency fee situation, the plaintiff's lawyers are entitled to a previously agreed upon fraction of any awards made, instead of legal fees, and in effect are advancing funds to finance the litigation. Personal injury lawyers believe that, without such contingency agreements, injured parties generally could not afford the legal fees involved. Opponents of this system believe that the contingency fee basis increases the number of cases and the amount of each claim, and therefore the total cost of liability litigation.

SECTION 7.2. FIRE LOSSES AND LITIGATION

The fire losses which result in litigation fall into two classes: personal injury and death, and property and business losses.

The litigation involving personal injury and death covers the following items:

1. Compensation for loss of life, which may include the estimated loss of earnings to support a family.
2. Compensation for hospital costs and other medical expenses.
3. Compensation for estimated future costs of special care and support.
4. Compensation for pain and suffering.

The litigation involving property and business losses covers the following items:

1. Compensation for loss of the structure, which may range from the amount of insurance to full replacement cost.
2. Compensation for loss of contents of the structure, which may include the value of contents, not only on the date of loss, but even on a later date at which the contents could have been sold for higher prices had it not been for loss of structure and contents.
3. Compensation for losses due to business interruption.

SECTION 7.3. VULNERABILITY OF PLASTICS TO FIRE LITIGATION

The factors affecting the relative vulnerability of a particular plastic to fire litigation are the following:

1. **Overall availability.** Large-volume commodity plastics are more likely to be involved in a fire than small-volume specialty plastics. For example, polyvinyl chloride, 2,469,000 metric tons of which were sold in 1980, is more likely to be involved in a particular fire than polycarbonate, only 99,000 metric tons of which were sold in 1980, the probability on this basis being 25:1.

2. **Market share of a particular application.** Plastics which comprise a large portion of a particular application are more likely to be involved in a particular fire than plastics which have only a small market share. For example, wire and cable applications in 1980 used 177,000 metric tons of polyvinyl chloride and less than 13,000 metric tons of nylon. On this basis, a randomly selected fire involving wire and cable would have polyvinyl chloride over thirteen times more likely to be involved than nylon.

3. **Relative vulnerability of the application.** A specific plastic used in a low-risk application is less likely to be involved in a fire than the same plastic used in a high-risk application. For example, the 26,000 metric tons

of high density polyethylene used in gas pipe in 1980 would be more likely to be involved in a fire than the 27,000 metric tons of the same plastic used in water pipe.

4. **Relative amount contained in a single product item.** A product item containing a large quantity of plastic is more likely to play a significant role in a fire than a small product item. For example, the polyvinyl chloride in a pair of footwear would play a much less significant role in a fire than the polyvinyl chloride wallcovering on all four walls of the room.

5. **Relative flammability of the plastic.** A particular plastic formulated for greater fire resistance would be less likely to be involved in a fire than a less fire resistant formulation. For example, a polyurethane flexible foam cushion in furniture made to be sold in California would be expected to be less likely to be involved in a fire than a similar cushion in furniture made to be sold in states other than California.

6. **Relative smoke-producing characteristics of the plastic.** A plastic material which tends to produce dense smoke would be more likely to be involved in litigation than one which tends to produce relatively little smoke. For example, in a fire in which dense black smoke was a major factor, polyvinyl chloride or ABS, if they could be identified among the materials involved, would be more likely to be blamed than polypropylene or phenolic.

7. **Relative toxicity of off-gases.** A plastic which has been extensively studied, and published, for toxicity of off-gases would be more likely to be involved in litigation than one which has received much less attention. For example, in a fire in which acrid, toxic fumes or post-fire difficulties were reported by firefighters, polyvinyl chloride, if it could be identified among the materials involved, would be more likely to be blamed than polyethylene.

8. **Relative size and sophistication of applicators.** A plastic which can be installed by an applicator with relatively little investment or training would be more likely to be involved in fire litigation than one which can be installed only with substantial investment and significant training. For example, many fire cases involving rigid polyurethane foam resulted from one-of-a-kind applications and relatively small applications by small contractors with minimal spray equipment and minimal training, while engineering plastics which require substantial investments in molding or extrusion equipment have had relatively little involvement in fire litigation.

9. **Relative control by manufacturer over ultimate use.** A manufacturer

which has knowledge and some measure of control over the ultimate application would be less likely to be involved in fire litigation than one which has no knowledge of intended use once the product has been delivered to a distributor. For example, neoprene flexible foam, which tends to be used only in well-defined applications such as mass-transit seating and shipboard mattresses, has been much less involved in fire litigation than polyurethane flexible foam, which is sold as slabs, blocks and even scrap by small outlets to any customer with few if any questions asked.

CHAPTER 8

Commercial Fire, Smoke, and Smolder Retardants

For many plastic materials, the most feasible method of improving their resistance to fire is the incorporation of commercially available retardants. In previous years, these additives were considered primarily fire retardants; many are now viewed as smoke retardants and smolder retardants because smoke and smoldering have become the fire response characteristics of concern in some applications.

Some materials used as fire retardants have been found to increase smoke, some other materials used as fire retardants have been found to increase smoldering, and yet other materials have been found to be effective with regard to two or even all three characteristics. Their extent of effectiveness is very much a function of the material and the application involved, and it is not possible at this time to classify retardants on the basis of the characteristics on which they have the most desirable impact.

The materials listed in Table 8.1 represent a broad spectrum of available retardants. The listing was prepared from information provided by nineteen manufacturers, and is not intended to be a complete listing of all available commercial products.

Table 8.1. Commercial Fire, Smoke, and Smolder Retardants

A & S CORPORATION		
Ammonium Sulfamate		
ALUCHEM INC.		
Ground hydrated aluminas		
AC-400K and AC-401		
AC-410K and AC-411		
AC-420K and AC-421		
AC-430K and AC-431		
AC-440K and AC-450K		
AC-460K and AC-470K		
AC-714K and AC-722K		
AC-740K		
DIAMOND SHAMROCK CORPORATION		
Chlorowax 20	chlorinated paraffin	22% Cl
Chlorowax 40	chlorinated paraffin	43% Cl
Chlorowax 45	chlorinated paraffin	45% Cl
Chlorowax 50	chlorinated paraffin	48% Cl
Chlorowax 50HV	chlorinated paraffin	50% Cl
Chlorowax 500C	chlorinated paraffin	59% Cl
Chlorowax 65	chlorinated paraffin	65% Cl
Chlorowax 70	chlorinated paraffin	69% Cl
Chlorowax 70L	chlorinated paraffin	69% Cl
Chlorowax 70S	chlorinated paraffin	69% Cl
Chlorowax 7435	chlorianted paraffin	47% Cl
Chlorowax 100	chlorinated paraffin	40% Cl
Chlorowax LV	chlorinated paraffin	39% Cl
Diablo 700X	chlorinated paraffin	70% Cl
Delvet 65	Chlorowax 70 in water	
DOVER CHEMICAL CORPORATION		
Chloroflo 40	chlorinated paraffin	38% Cl
Chloroflo 42	chlorinated paraffin	39% Cl
Paroil 10	chlorinated paraffin	40% Cl
Paroil 140	chlorinated paraffin	42% Cl
Paroil 142A	chlorinated paraffin	45.5% Cl
Paroil 145A	chlorinated paraffin	46.5% Cl
Paroil 150A	chlorinated paraffin	50% Cl
Paroil 152	chlorinated paraffin	51% Cl
Paroil 1550	chlorinated paraffin	54.5% Cl
Paroil 1160	chlorinated paraffin	59% Cl
Paroil 1650	chlorinated paraffin	62% Cl
Paroil 170	chlorinated paraffin	70% Cl

Continued

Table 8.1. Commercial Fire, Smoke, and Smolder Retardants

DOVER CHEMICAL CORPORATION (continued)		
Paroil 170LV	chlorinated paraffin	67% Cl
Paroil 170HV	chlorinated paraffin	70% Cl
Paroil 170T	chlorinated paraffin	70% Cl
Paroil 170-7	chlorinated paraffin	72% Cl
Paroil 170-8	chlorinated paraffin	72% Cl
Chlorovis 150A	chlorinated paraffin	44–45% Cl
Chlorez 700	chlorinated paraffin	70% Cl
Chlorez 700HMP	chlorinated paraffin	70% Cl
Chlorez 700S	chlorinated paraffin	70% Cl
Chlorez 760	chlorinated paraffin	74% Cl
DOW CHEMICAL COMPANY		
FR-300-BA	decabromodiphenyl oxide	83% Br
FR-1138	dibromoneopentyl glycol	61% Br
FR-3675, XNS-500S4.20		36% Br
XNS-50097		42.7% Br
EMERY INDUSTRIES		
Emery 9331 DBP	dibromophenol	63.4% Br
Emery 9332 TBP	tribromophenol	72.4% Br
Emery 9336 DBNPG	dibromoneopentyl glycol	61% Br
Emery 9345 TBX	tetrabromoxylene	75% Br
Emery 9350 TBBA	tetrabromobisphenol A	58% Br
Emery 9353 EOTBBA	tetrabromobisphenol A di-2-hydroxyethyl ether	48.5% Br
FERRO CORPORATION, KEIL CHEMICAL DIVISION		
Bromoklor 50	halogenated aliphatic liquid	30% Br, 20% Cl
Bromoklor 70	halogenated aliphatic liquid	35% Br, 35% Cl
HARSHAW CHEMICAL COMPANY		
HFR-131	ammonium fluoborate and antimony trioxide	
HFR-201	30% pentavalent antimony sol, alkalized	
HFR-301X	pentavalent antimony oxide, alkalized	
HUMPHREY CHEMICAL CORPORATION		
ZB-112	zinc borate	45% ZnO, 35% B_2O_3
ZB-112 R	zinc borate	47% ZnO, 34% B_2O_3
ZB-237	zinc borate	33% ZnO, 41% B_2O_3
ZB-325	zinc borate	52% ZnO, 29% B_2O_3
ISOCHEM RESINS COMPANY		
Isobor	zinc boro phosphate	
Isochem I-599	halogen phosphate	60% halogen

(Continued)

Table 8.1. Commercial Fire, Smoke, and Smolder Retardants (Continued)

M & T CHEMICALS		
Thermoguard S	antimony oxide	83.0% Sb
Thermoguard L	antimony oxide	83.0% Sb
Thermoguard S-800	antimony oxide	80.0% Sb
Thermoguard S-711	antimony oxide	70.0% Sb
Thermoguard FR	antimony oxide	61.7% Sb
Thermoguard CPA	antimony oxide	41.0% Sb
MONSANTO COMPANY		
Santicizer 141	2-ethylhexyl diphenyl phosphate	
Santicizer 143	modified triaryl phosphate ester	
Santicizer 148	isodecyl diphenyl phosphate	
Santicizer 154	t-butylphenyl diphenyl phosphate	
Phosgard 2XC-20	chloro-phosphate ester	
Phosgard C-22-R	organo-phosphorus polymer	27% Cl, 15% P
Triphenyl phosphate		
SAMINCORP		
Antimony oxide		
SANDOZ INC.		
Sandoflam 5060		
Sandoflam 5070		
SHERWIN WILLIAMS COMPANY		
KemGard 425		
KemGard 911A		
KemGard 911B		
KemGard 911C		
SS-121		
STAUFFER CHEMICAL COMPANY		
Fyrol EFF	oligomeric chloroalkyl phosphate	11.5% P, 32.5% Cl
Fyrol FR-2	tri(B,B′-dichloroisopropyl)phosphate	7.1% P, 49.0% Cl
Fyrol PCF	tri(B-chloropropyl)phosphate	9.46% P, 32.5% Cl
Fyrol CEF	tri(B-chloroethyl)phosphate	10.8% P, 36.7% Cl
Fyrol DMMP	dimethyl methylphosphonate	25% P
Fyrol 6	diethyl N,N-bit(2-hydroxyethyl)amino-methylphosphonate	12.4% P
Fyrol 99	chlorinated oligomeric phosphate ester	14.0% P, 26.0% Cl
Fyrol 51		20.5% P
Fyrol 58		17% P
Lindol	tricresyl phosphate	
Phosflex 41P	isopropylated triaryl phosphate ester	7.9% P
Phosflex 61B	butylated triaryl phosphate ester	8.0% P
Phosflex 71B	butylated triphenyl phosphate ester	8.5% P

(Continued)

Table 8.1. Commercial Fire, Smoke, and Smolder Retardants (Continued)

UNITED STATES BORAX AND CHEMICAL COMPANY		
Firebrake ZB	zinc borate	
VELSICOL CHEMICAL CORPORATION		
Firemaster PHT4	tetrabromophthalic anhydride	68.2% Br
Firemaster 100	hexabromocyclododecane	74.7% Br
Firemaster CA	chlorendic anhydride	54.5% Cl
Firemaster 680	1,2-bis(2,4,6-tribromophenoxy) ethane	70.0% Br
Firemaster 935	polydibromophenyleneoxide	63.0-65.5% Br
Firemaster BP4A	tetrabromobisphenol-A	57.7% Br
2,4,6-tribromophenol		72% Br
WITCO CHEMICAL CORPORATION, PEARSALL CHEMICAL DIVISION		
Fyarestor 100	bromochlorinated paraffin	40% Cl, 20% Br
Fyarestor 330		55% Br, 8% P

(Concluded)